00-CTS-14ET

A NOVEL MEMBRANE PROCESS FOR AUTOTROPHIC DENITRIFICATION

by:

Dr. Kauser Jahan
Rowan University

2003

Water Environment
Research Foundation

The Water Environment Research Foundation, a not-for-profit organization, funds and manages water quality research for its subscribers through a diverse public-private partnership between municipal utilities, corporations, academia, industry, and the federal government. WERF subscribers include municipal and regional water and wastewater utilities, industrial corporations, environmental engineering firms, and others that share a commitment to cost-effective water quality solutions. WERF is dedicated to advancing science and technology addressing water quality issues as they impact water resources, the atmosphere, the lands, and quality of life.

For more information, contact:
Water Environment Research Foundation
635 Slaters Lane
Alexandria, VA 22314-
Tel: (703) 684-2470
Fax: (703) 299-0742
www.werf.org
werf@werf.org

This report was published by the following organization. For non-subscriber sales information, contact:
IWA Publishing
Alliance House, 12 Caxton Street
London SW1H 0QS, United Kingdom
Tel: +44 (0) 20 7654 5500
Fax: +44 (0) 20 7654 5555
www.iwapublishing.com
publications@iwap.co.uk

Library of Congress Catalog Card Number: 2003103356

WERF ISBN 978-1-893664-65-4 IWAP ISBN: 978-1-84339-643-7

Rowan University

ACKNOWLEDGMENTS

The project team would like to acknowledge Rowan University and Dr. Tariq Ahmed of TAMS Consultants for their outstanding support of this project.

Report Preparation

Principal Investigator:
Dr. Kauser Jahan, Ph.D., P.E.
Rowan University

Project Team:
Agnieszka Pierkiel, M.S.
Claire Steager, M.S.
Rowan University

Project Subcommittee

Research Council Liaison:
Dr. Bruce E. Rittmann, Ph.D.
Northwestern University

Bernard Sawyer
Metropolitan Water Reclamation District of Greater Chicago

Water Environment Research Foundation Staff

Research Program Director: Jeff Moeller, P.E.
Project Manager: Mary E. Strawn

ABSTRACT AND BENEFITS

Abstract

Biological denitrification by heterotrophic bacteria is common in the wastewater industry in the United States and in drinking water processing in Europe. To facilitate heterotrophic denitrification, organic compounds such as methanol, ethanol, and acetic acid are added to provide a carbon source for the bacteria. The resulting organic carbon residual may create problems with chlorination. The addition of these carbon compounds is expensive and results in more sludge being produced.

This study focused on the use of autotrophic, hydrogen-oxidizing bacteria for denitrification. The method transfers hydrogen gas to solution via microporous hollow fiber membranes. Typically, gases are supplied to a system using conventional bubble diffusers. The conventional bubble aeration system has low gas transfer efficiency; consequently, the cost of dissolving the required amount of gas is very high. For this study, microporous hollow fiber membranes were used to supply hydrogen gas to hydrogen-oxidizing, autotrophic bacteria enriched from mixed cultures obtained from wastewater sludge.

The use of hollow fiber membrane modules appears to be a viable technology for transferring hydrogen gas to water. The experimental results described in this report provide valuable information for the water/wastewater community about pilot- and full-scale membrane processes for autotrophic denitrification.

Benefits

- Using an inorganic carbon source for denitrification could reduce costs significantly over processes that use an organic C source such as methanol, ethanol or acetic acid.
- Use of inorganic C source prevents the potential for TOC breakthrough in the plant effluent that is of concern with organic C addition.
- Using H2 as the e-donor is advantageous because the residual doesn't have the potential interfere with subsequent treatment (unlike sulfur, ferrous or chloric compounds).
- The "dead end" configuration for membrane module design yields 100% gas transfer efficiency into the bulk liquid, at the same time preventing hydrogen bubbling and therefore the stripping of dissolved volatile compounds from the liquid.
- Fluidization of membrane fibers may reduce biological and chemical fouling.
- The best nitrate removal efficiency per unit area can be achieved at an HRT between 1.75 and 3.25 h.
- At HRTs of 3.25 h or more, hydrogen transfer was greater than the stoichiometric requirement, meaning H2 gas is not limiting.

- Continuous flow studies indicated that a stable biofilm can be developed in a packed bed reactor to remove nitrate using H2 as the electron donor.

Keywords: Autotrophic denitrification, hollow fiber membranes, hydrogen oxidizing bacteria, hydrogen dissolution, nitrate.

CONTENTS

LIST OF TABLES

LIST OF FIGURES

LIST OF ACRONYMS

AWWA	American Water Works Association
BOD	biochemical oxygen demand
CMFR	completely mixed flow reactor
DO	dissolved oxygen
DOC	dissolved organic carbon
ED	electrodialysis
HFM	hollow fiber membrane
HFMBR	hollow fiber membrane bioreactor
HRT	hydraulic retention time
ID	inside diameter
IX	ion exchange
MCL	mean contaminant level
MLSS	mixed liquor suspended solids
MLVSS	mixed liquor volatile suspended solids
OTE	oxygen transfer efficiency
PDMS	polydimethylsiloxane
PTFE	polytetrafluoroethylene
PVC	polyvinylchloride
RO	reverse osmosis
TOC	total organic carbon
EPA	U.S. Environmental Protection Agency
USGS	U.S. Geological Survey
VOC	volatile organic carbon

EXECUTIVE SUMMARY

This project focused on the novel use of hollow fiber membranes in biological denitrification using hydrogen-oxidizing bacteria. In the United States, ion exchange (IX) and biological denitrification are the most widely used technologies for drinking water and wastewater, respectively. The IX process is associated with high operating costs because of the presence of competing ions and regenerant disposal problems. The heterotrophic denitrification process requires addition of an external organic carbon source, such as methanol, ethanol, or acetate. If organic carbon addition is not controlled accurately in the bioreactor, the residual carbon in the finished water is likely to interfere with the chlorination process.

Autotrophic denitrification is also a biological process that reduces nitrate to nitrogen gas using autotrophic, hydrogen-oxidizing bacteria. Carbon dioxide or bicarbonate can be used as the inorganic carbon source for the process. Nitrate acts as the electron acceptor whereas hydrogen gas is the electron donor. Typically, gases are supplied to a system using conventional bubble diffusers. The conventional bubble aeration system has low gas transfer efficiency; consequently, the cost of dissolving the required amount of gas is very high. If, however, hydrogen can be transferred to water with 100% gas transfer efficiency, preliminary calculations indicate that autotrophic denitrification may be very cost-effective compared to heterotrophic denitrification.

For this study, microporous hollow fiber membranes were used to supply hydrogen gas to hydrogen-oxidizing, autotrophic bacteria. Laboratory-scale membrane modules were constructed, and mass transfer studies were carried out to develop the design correlations for hydrogen gas transfer. A mixed culture was obtained and acclimated for batch denitrification studies. Sodium carbonate and carbon dioxide were used to deliver inorganic carbon. Bench-scale, continuous-flow biofilm reactors containing plastic media were operated to remove nitrate from water. The required hydrogen gas was supplied at a constant rate via gas transfer modules containing sealed-end microporous hollow fiber membranes. The reactors were optimized for removal of nitrate and nitrite by varying recycle ratios and hydraulic retention times.

Experimental results indicated the presence of hydrogen oxidizing denitrifiers in wastewater treatment plants. A mixed culture capable of using hydrogen as an electron donor was easily developed. Adequate pH control was possible, and the pH averaged around 6.95. Gas transfer studies indicated that hydrogen transfer was primarily controlled by liquid film diffusion. The following mass transfer correlation can be used to design membrane modules for hydrogen gas transfer to water:

$$Sh = 2.68\ (Re\,\mathrm{de/L})^{1.02}\ Sc^{0.33}$$

where *Sh* is the Sherwood number, *Re* is the Reynolds number, de is the equivalent diameter of the membrane module, L is the length of the fibers, and *Sc* is the Schmidt number.

Continuous flow studies indicated that a stable biofilm can be developed in a packed bed reactor to remove nitrate using hydrogen as the electron donor. Hydrogen gas was successfully delivered to the reactor via the hollow fiber membrane gas transfer module without fouling. Hydrogen concentrations indicated that the system did not experience hydrogen limitations at retention times of 3.25 hours or greater.

The membrane hydrogen transfer technology appears to be a viable technology for transferring hydrogen gas to water. The results reported here may serve as a basis for pilot- and full-scale studies of membrane processes for autotrophic denitrification.

CHAPTER 1.0

INTRODUCTION

1.1 Background

1.1.1 Nitrate in the Environment

Nitrate is one of the most common groundwater contaminants in the United States and elsewhere (Korom, 1992; Freeze and Cherry, 1979; Gayle et al., 1989; Kapoor and Viraraghavan, 1997). It is a thermodynamically stable and highly soluble nitrogen species that is easily transported and accumulated in groundwater systems. These properties, coupled with increased anthropogenic releases of nitrogen-containing compounds, have resulted in elevated nitrate concentrations in United States and European ground- and surface waters.

Nitrate contamination originates from point and nonpoint sources. The majority of point source contributions are attributed to domestic and industrial wastewater discharges. Contamination of drinking water sources with nitrate may result from both artificial and organic overfertilization in agriculture, human and animal waste disposal, and discharge of the wastewater from food processing, explosives manufacturing, NOX absorption in air stripping, and recovery of nuclear fuels (Pekdemir et al., 1998). Nonpoint sources have a larger impact and are associated with agricultural practices, fisheries, poultry operations, livestock feeding, and residential septic tank effluents (Hallberg, 1989). Many drinking water aquifers have nitrate concentrations approaching or already above the maximum contaminant level (MCL) of 10 mg/L NO_3^--N established by the U.S. Environmental Protection Agency (EPA, 1993). An investigation conducted by the U.S. Geological Survey (AWWA, 1995) analyzed nutrient concentrations of 12,000 groundwater and 22,000 surface water samples between 1970 and 1992. The results of the survey indicated that nitrate levels in 20% of shallow private wells in U.S. farming areas violate the federal drinking water standard and that 1 percent of public water supply wells exceed the limit. Surface waters down gradient of agricultural areas had elevated nitrate concentrations, but rarely exceeded the limit. Of the 3351 private wells surveyed, 9% had nitrate concentrations exceeding the 10 mg/L standard.

1.1.2 Health and Environmental Consequences

Incidence of methemoglobinemia is related to ingestion of nitrate contaminated water (Shuvai and Gruener, 1977). Methemoglobinemia or "blue baby syndrome" occurs in infants under 6 months old and can be life threatening without immediate medical attention. N-nitroso compounds are also formed from nitrates and are known to be carcinogenic. Nitrites are intermediates in nitrate biodegradation and are toxic to aquatic organizms (Metcalf and Eddy, 1991). Nitrate disposal in aquatic systems also accelerates the onset of eutrophication (Metcalf and Eddy, 1991).

Health problems related to consumption of nitrate-contaminated drinking water have led to regulation by EPA. The nitrate mean concentration standards (MCLs) require limits of less than 10 ppm NO_3^--N/L and 1ppm NO_2^--N/L in drinking water. The European standards are stricter, allowing the concentration of 0.03 ppm NO_2^--N/L (Urbain et al., 1996). A recommended level for nitrate is 5.6 ppm NO_3^--N/L for the European Community (Kapoor and Viraraghvan, 1997). To comply with these standards, municipal water suppliers are faced with the need for efficient and cost effective nitrate removal processes.

1.1.3 Nitrate Removal

Nitrate removal, or denitrification, is usually accomplished through physicochemical or biological processes. Several methods are available for removing nitrate from water and wastewater (Weber and DiGiano, 1996; Eckenfelder, 1989). Separation methods used in denitrification include ion exchange (IX), reverse osmosis (RO), activated-carbon adsorption, and electrodialysis (ED). Advantages include immediate nitrate removal at process startup, simple process control, reliability, and ease of monitoring. For drinking water, IX is currently the predominant method for removing nitrate in the United States. Even with the recent development of nitrate selective resins, maintenance and operating costs are still high, and the production of concentrated waste brines creates additional disposal problems. The biological denitrification of spent ion exchange regenerant is an effective alternative to continuous disposal (Viraraghavan and Rao, 1990; van der Hoek et al., 1988; Clifford and Liu, 1993). The long-term operational problem with an IX/denitrification process is that anion exchange resins are susceptible to significant organic fouling.

In European countries, biological denitrification of drinking water using heterotrophic microorganisms held in fixed bed reactors is more common (Delanghe et al., 1994a and b; Lazarova et al., 1992). These systems typically use methanol, ethanol, or acetic acid as the organic carbon source. The residual organics, if not removed effectively, can increase the chlorine demand of the water or react with chlorine and form carcinogenic disinfection byproducts (DBPs). The unconsumed organic matter contributes to the biological instability of the drinking water (Rittmann and Huck, 1989). The adverse effects associated with biological instability in the distribution system are: (a) increased microbial growth; (b) production of taste and odor generating compounds; and (c) increased corrosion of pipes (Rittmann and Snoeyink, 1984). These risks have limited heterotrophic denitrification to wastewater applications in the United States, where chlorine is the predominant means for disinfecting potable water. In Europe, disinfection of drinking water is being accomplished more by ozone or ultraviolet (UV) radiation.

1.1.4 Heterotrophic Denitrification

Biological denitrification is a microbial process that uses microorganisms to convert nitrate to nitrogen gas. Bacteria use nitrate as the terminal electron acceptor in their respiratory processes in the presence of low concentrations of oxygen. Denitrification consists of a sequence of enzymatic reactions leading to the evolution of nitrogen gas. The process may involve the formation of the following intermediates (Szekers et al., 2001):

$$\mathbf{NO_3^- \rightarrow NO_2^- \rightarrow NO \rightarrow N_2O \rightarrow N_2}$$

Biological dentrification can be heterotrophic or autotrophic. Heterotrophic denitrification is the most common method of removing nitrogen from municipal and industrial wastewaters (Metcalf & Eddy, 1991). The denitrification reactors follow nitrification processes that convert organic nitrogen and ammonia to nitrate. The denitrification reactors can be either suspended-growth or attached-growth, with the latter being favored because it minimizes the space required. The disadvantages of heterotrophic denitrification are the costs incurred for adding an external organic carbon source for cell synthesis and the posttreatment needed to remove the residual organic contaminant (precursor for DBPs) and biomass before disinfection and final discharge.

1.1.5 Autotrophic Denitrification

For autotrophic organisms, no organic carbon source is required; rather, carbon dioxide or bicarbonate is used for cell synthesis (Brezonik, 1977). For the energy source, autotrophic organisms require hydrogen or reduced-sulfur compounds. There are four groups of autotrophic denitrifiers (Mateju *et al.*, 1992): hydrogen-oxidizing bacteria, reduced-sulfur-oxidizing bacteria, ferrous-iron-oxidizing bacteria, and chloric-compound-oxidizing bacteria. Hydrogen-oxidizing bacteria reduce nitrate or nitrite using hydrogen as an electron donor and energy source. Autotrophic denitrification with hydrogen-oxidizing bacteria is preferred for drinking water treatment because: (1) the residual hydrogen that remains in water is harmless and does not interfere with subsequent water treatment (Rittmann and Huck, 1989); (2) only inorganic carbon is used and, therefore, there are no organic residues for further treatment; and (3) reaction byproducts are harmless.

The following stoichiometric relationships have been proposed for autotrophic denitrification with thiosulfate (Claus and Kutzner, 1985) and sulfide (Barrenstein et al., 1986; van Houten et al., 1994) serving as the electron donors:

$$5S_2O_3^{=} + 8NO_3^- + H_2O \rightarrow 4N_2 + 10SO_4^{=} + 2H^+ \qquad (1.1.5\text{-}1)$$
$$5S^{=} + 8NO_3^- + 8H^+ \rightarrow 5SO_4^{=} + 4N_2 + 4H_2O \qquad (1.1.5\text{-}2)$$

Autotrophic organisms, e.g., *Micrococcus denitrificans*, reduce nitrate to nitrite and subsequently to nitrogen gas while oxidizing hydrogen gas to water. The reaction proceeds as follows (Kurt et al., 1987):

$$2NO_3^- + 2H_2 \rightarrow 2\,NO_2^- + 2H_2O \qquad (1.1.5\text{-}3)$$
$$2NO_2^- + 3H_2 \rightarrow N_2 + 2H_2O + 2OH^- \qquad (1.1.5\text{-}4)$$

Combining reactions 1.1.5-3 and 1.1.5-4 gives the overall reaction:

$$2NO_3^- + 5H_2 \rightarrow N_2 + 4H_2O + 2OH^- \qquad (1.1.5\text{-}5)$$

From the overall reaction given above, 2 moles of OH^- are produced for every 2 moles of NO_3^- reduced; thus alkaline conditions (high pH) are possible and NO_2^- accumulation can be expected. To prevent pH shifts and nitrite accumulation, CO_2 or HCO_3^- can be added to buffer the system, as well as serve as the inorganic carbon source for cell synthesis. From the above equation, theoretically 0.35 mg H_2/mg NO_3^--N is required for complete reduction of nitrate to nitrogen gas. Experimental values were 0.38 mg H_2/mg NO_3^--N in a bench-scale reactor (Kurt et al., 1987) and 0.40 mg H_2/mg NO_3^--N at a full-scale plant in Germany (Gross et al., 1986). The higher values probably reflect that some dissolved oxygen was reduced by hydrogen oxidation.

Recently, interest has escalated to develop systems that use autotrophic bacteria for biological drinking water denitrification. Novel configurations are being developed which incorporate different process configurations.

1.1.6 Autotrophic Denitrification in Fixed Biofilm Reactors

Kurt et al. (1987) conducted extensive studies using autotrophic hydrogen-oxidizing, denitrifying bacteria in a bench-scale fluidized bed reactor. Nitrite tended to accumulate in batch tests, but complete denitrification occurred in continuous flow experiments with sufficiently long residence times. A residence time of 4.5 h was required for complete denitrification of water containing 25 mg/L NO_3^-–N. The optimum pH was found to be approximately 7.5, and the pH increased under unbuffered conditions. If the pH was allowed to rise near 9.0, nitrite tended to accumulate. The denitrification kinetics was modeled as a two-step process using a double Monod saturation function:

$$\mu_{NO3/NO2} = \mu_{m(NO3/NO2)} \frac{C_{(NO3/NO2)}}{K_{S(NO3/NO2)} + C_{(NO3/NO2)}} * \frac{C_{H2}}{K_{S(H2)} + C_{H2}} \qquad (1.1.6\text{-}1)$$

where μ and μ_m are the specific and maximum specific growth rates of autotrophic organisms, respectively, and C and K_S are the liquid phase substrate concentration and half-velocity constant, respectively. Nitrate and hydrogen were assumed to be limiting substrates in the first denitrification step, and nitrite and hydrogen were assumed to be limiting in the second step as given in Equations 1.1.5-3 and 1.1.5-4. The denitrification rate was shown to be more strongly dependent on the nitrate concentration than on hydrogen. Although hydrogen is only slightly soluble in water (1.6 mg/L at 20°C), the Monod half-velocity constant for hydrogen was reported to be less than 1% of saturation. Therefore, it is evident that the liquid phase hydrogen concentration has very little effect except at low nitrate concentrations.

Gross et al. (1986) and Gross and Treutler (1986) reported on the startup and performance of a commercial-scale biological denitrification plant utilizing hydrogen. The technology was developed by Sulzer Water and Wastewater Treatment with the trade name DENITROPUR. The plant reached full capacity (50 m^3/h) in approximately 3 months. The plant design incorporated indirect hydrogen saturation, phosphate addition, four packed-bed reactors in series, postaeration, flocculant addition, filtration, and ultraviolet disinfection. Carbon dioxide was added as an inorganic carbon source and to buffer against an alkaline pH shift. At the operating temperature of 10.5°C, the organism growth rate varied from 0.1 to 0.3 per day. The sludge production was reported to be approximately 0.2 kg per kg nitrogen removed on a dry-weight basis. Residence times of 1 to 2 hours were required to remove 50 mg/L NO_3^-.

Benedict et al. (1998) investigated the effect of reactor packing media on autotrophic denitrification with hydrogen-oxidizing bacteria. Hydrogen and carbon dioxide were sparged into the feed water. The bench-top biofilm reactors were packed with granular activated carbon, sand, pumice, volcanic rock, and polymer rings. A mixed culture was obtained from a heterotrophic denitrification process in a water treatment facility. Nitrate and nitrite reduction was monitored daily at different hydraulic retention times. The highest removal efficiency with respect to surface area (770.4 mg NO_3^--N/day) was achieved in the polymer ring bioreactor with an 8-hour hydraulic retention time. The performance of the polymeric rings medium was found to be superior to other investigated media in the range of 2 to 8 hours retention times. Nitrite reduction decreased with decreasing hydraulic retention time.

Long-term performance of an autotrophic denitrification process was evaluated by Szekeres et al. (2001). The process consisted of two steps: (1) the water to be treated was enriched with hydrogen in the cathodic chamber of an electrochemical cell and (2) denitrified in a fixed-bed biofilm reactor. In the electrochemical cell, Nafion 417 cation-exchange membrane separated the two electrodes. Water was split by electrolysis into diatomic hydrogen and OH^- ions at the cathode. The feed solution — consisting of tap water with nitrate, carbonate, and phosphorus — was thus hydrogen enriched in the cathodic chamber. As a result of the net reaction of denitrification, cathodic and anodic reaction, one mole of protons passed through the cation exchange membrane for each mole of nitrate. The reactor was a packed bed of granulated activated carbon in an upflow configuration.

The bioreactor was inoculated with a mixture of four denitrifying strains isolated from a previous reactor. The original inoculum was an enrichment culture from sediment from an oxidation pond treating domestic wastewater. The system was operated in a continuous mode for a period of 1 year. The feed concentration was kept constant at 21 mg N/L, and water velocities ranged from 0.04 m/h to 0.2 m/h with current densities from 40 mA to 100 mA. The rates of denitrification increased with water velocity at constant current of 80 mA until it reached 0.15 m/h when approximately 0.23 kg N/m^3day was removed with an effluent nitrite concentration below 1 mg N/L. The peak nitrate removal rate of 0.19 kg N/m^3day occurred at 70 mA in the runs at constant water velocity of 0.11 m/h. Denitrification rates up to 0.25 kg N/m^3day were obtained at the hydraulic residence time of 1 h.

1.1.7 Single-Step Denitrification in Fixed-Bed Reactors

Dries et al. (1988) tested a two-column system with removal of nitrate in the first column using polyurethane as support medium and removal of excess hydrogen and oxidation of residual nitrite to nitrate in the second column. Water flowed downward in the denitrification column while hydrogen entered from the bottom. The water then passed through the second column in an upflow mode. Denitrification rates of 0.5 kg N/m^3day were obtained at 20°C.

Electrodes were embedded in laboratory-scale sandy aquifers, and hydrogen gas was produced through the electrolysis of water for enhancing autotrophic denitrification (Sakakibara and Kusaka, 1999). Groundwater containing 20 mg N/L of $NaNO_3$ and no electron donor was fed continuously to the synthetic aquifer. The denitrification rate varied depending on applied current. Electrical power requirement was 0.03–0.09 kWh/g N. Results showed that when carbon electrodes were embedded in a direction perpendicular to groundwater flow, complete and stable denitrification was achieved over 1 year.

Trouve and Chazal (1999) removed nitrate from groundwater using an environmental strain of *Thiobacillus denitrificans.* A pilot plant in fixed bed conditions filled with a pozzolana/neutral G mixture was used. The feed solution contained nitrate and thiosulfate in tap water at 10°C. An optimum volumetric nitrate loading rate of 1.5 kg NO_3^--N/m^3-day was determined for which 100% denitrification yield was reached. The culture exhibited good denitrification abilities at low temperatures: the mean nitrate removal velocity was twice as great at 10°C than at 20°C.

Immobilized systems can also be used for autotrophic denitrification in column configurations. *Ralstonia eutropha* (formerly *Alcaligenes eutrophus)*, a hydrogen-oxidizing bacterium, was immobilized in polyacrylamide and alginate copolymer beads, which were then used as media in a fluidized bed reactor (Chang, 1999). The total reactor volume was 1 L, and 20% of the reactor volume was occupied by the immobilized beads. The reactor operated in continuous and batch modes. Feed contained nitrate and bicarbonate in deionized water, and the hydraulic retention time was 400 min. The total nitrogen removal rate in a continuous test increased with operation time and reached a maximum rate of 0.6–0.7 kg N/m^{3}day after 6 days. Nitrite reductase was inhibited when the dissolved hydrogen concentration fell below 0.2 mg/L, whereas nitrate reductase was inhibited at a concentration below 0.1 mg/L. A feed containing nitrate, bicarbonate, and phosphate in tap water was also investigated. Phosphate concentration affected denitrification, especially in the accumulation of nitrite: as the concentration of phosphate increased, the accumulation of nitrite decreased. The bacteria were found to adapt to a shock nitrate loading. Although one full-scale system is being used in Europe, the process did not receive much attention because of the following reasons:

- lack of an effective hydrogen gas dissolution device; and
- safety issues associated with the use of hydrogen.

Both of these concerns can be resolved using a membrane gas transfer device (Semmens, 1991). The membrane system can be used to transfer hydrogen with 100% gas transfer efficiency, and the system's bubble-free operation prevents the release and waste of hydrogen that can lead to explosive conditions in confined spaces.

1.1.8 Membrane Applications in Autotrophic Denitrification

Membrane technology was used for simultaneous hydrogen enrichment and denitrification in a single membrane bioreactor (Ergas and Reuss, 2001). Autotrophic denitrification using hydrogen-oxidizing bacteria was carried out in a hollow-fiber membrane biofilm reactor (HFMBR). More recently, hydrogen gas was transferred through hollow fiber membranes directly to a bacterial culture to accomplish biological denitrification of drinking water (Ergas and Reuss, 2001). A shell-and-tube hollow fiber membrane module was used in a flow-through configuration. A hydrogen-oxidizing denitrifying biofilm was attached to the outside of the hollow fiber membranes. Water contaminated with nitrate was recycled through the shell side. Nitrate diffused from the bulk fluid to the bacterial culture, which converted it to nitrogen gas effectively denitrifying the water. Hydrogen gas was pumped through the lumen of the fibers and diffused through the membrane to the bacterial culture.

A bench-scale HFMBR utilized 2400 polypropylene hollow fibers with an inner diameter of 200 μm, an outside diameter of 250 μm, and 0.05-μm pore size. A hydrogen-oxidizing denitrifying culture was enriched from a wastewater seed. The nitrate loading rate was gradually increased over a 3-month period following a 70-day startup period. The nitrate utilization rate reached 770 g NO_3^--N/m^{3}day at an influent nitrate concentration of 145 mg NO_3^--N/L. The hydraulic residence time was 4.1 h. Trial runs with contaminated water from the Cape Cod aquifer resulted in an increase in effluent water dissolved organic carbon (DOC) concentration and turbidity.

HFMBR technology was first applied to hydrogenotrophic denitrification by Lee and Rittmann (2000). The shell-and-tube configuration was used. Hollow fiber membranes used were composite membranes manufactured by Mitsubishi Rayon (Model MHF 200TL). The outer and inner layers of the fiber membrane wall were microporous polyethylene. A 1-μm-thick layer of nonporous polyurethane was sandwiched between the inner and the outer layers. The nonporous layer allows for high pressure use without premature bubble formation. The modules were composed of 83 hollow fibers with total gas transfer area of 750 cm^2. The total reactor volume was 420 mL. The hollow fiber membrane bundle was housed in a 1.5-cm ID PVC tube in an upflow configuration. The hollow fibers were sealed on one end and pressurized with hydrogen gas on the other end. Hydrogen gas supplied from a pressurized gas tank at 0.31 atm and 0.42 atm diffused through the membranes into the water stream flowing co-currently on the outside of the membranes.

The outer wall of each fiber membrane was coated with a biofilm. The organism was *Ralstonia eutropha* (formerly *Alcaligenes eutrophus)*, a hydrogen-oxidizing bacterium. The feed solution of tap water containing nitrate, bicarbonate, and 10-mM phosphate was pumped in the shell side of the module with a hydraulic retention time of 42 minutes. Nitrate diffused into the biofilm from the bulk liquid side of the biofilm, while hydrogen was delivered from the fiber side. Seventy-six percent nitrate removal was achieved at the hydrogen pressure of 0.31 atm, and 92% removal was achieved of 0.42 atm. High steady-state nitrate fluxes of 0.08 and 0.1 mg N/cm^{2}d were achieved at the dissolved hydrogen concentration of 0.009 and 0.07 mg H_2/L in the effluent. The biofilm thickness was 179 μm.

Hollow fiber membranes have also been used in other scenarios to achieve gas transfer to biofilms in denitrification. An HFMBR was used to treat 2,4,6-trinitrophenol as the sole carbon source. Pure oxygen was fed from the membranes to the biofilm attached to the outer surface of the membrane (Jimenez et al., 1998). Brindle et al. (1998) developed a denitrification process using a hollow fiber membrane reactor. A nitrifying biofilm was attached to the surface of oxygen-permeable hollow fiber membranes. Nitrogen removal of 98% from the nitrogen loading of 1.2 kg NH_4^--N/m^{2}day at the rate of 6.6 g NH_4^--N/m^{3}day was accomplished.

Membrane bioreactors effectively eliminate the need for posttreatment associated with biological denitrification. Because denitrification requires addition of bacteria, a carbon source, and buffering salts such as bicarbonate and phosphorus to the water to be treated, residual amounts of these chemicals raise the cost of posttreatment of denitrified water. If carbon dioxide is used as a carbon source, the treated water's pH rises again, adding to posttreatment costs.

Membrane bioreactors eliminate the problem of drinking water contamination by utilizing their separation capabilities along with gas transfer characteristics. Fuchs et al. (1997) separated the biological reaction zone and carbon supply from the raw water stream by a nitrate permeable membrane. Nitrate diffused from the water through the membrane into the biofilm while carbon and phosphate were transferred to the bacteria on the biofilm side of the membrane. In effect, the effluent was not contaminated with residual bacteria and chemicals because membranes separated the two solutions. Cellulose dialysis membrane tubing (7.5 mm ID) was used with a mixed heterotrophic culture. A denitrification rate of 1230 mg NO_3^--N/m^{2}d was achieved in this configuration. Even without posttreatment, the effluent met nearly all criteria for drinking water.

Membrane fluidized-bed technology combines the advantages of a fluidized bed (i.e., optimal mass transfer conditions combined with very high biomass concentration on fine-grained carrier particles with a very high volumetric surface) with the optimal bubble-free gas intake through permeation membranes. Hydrogen and carbon dioxide concentrations and pH are constant because of even distribution in the whole fluidized bed volume. Therefore, optimal conditions are available to microorganisms in the whole reactor resulting in very high specific denitrification rates.

The achievable reaction rate essentially depends on the kind and the quantity of the carrier particles and on the rate of the hydrogen intake. At limiting hydrogen rates or carrier limitation a maximum reaction rate is obstructed independently of nitrate load. A pilot plant was operated at retention times of about 15 min. Denitrification rates of 2.5 kg NO_3^--N/m^{3}d were obtained over longer periods for hydrogen-limited operation, which can be further increased by increasing the membrane surface or membrane permeability (Hoerner and Irmer, 1989). The membrane fluidized-bed technology is applied to autotrophic denitrification in Wurzen Waterworks (Germany).

Biologically treated wastewater was effectively denitrified in a fluidized bed using hydrogen as electron donor when the hydrogen was introduced through silicone rubber tubular membranes with specific area of 120 m^2/m^3 (Brautigam and Sekoulov, 1987). The denitrification rate was 60 g NO_3^--N/m^3-h at hydrogen partial pressure of 2 bar. Over time, biological growth on the membrane decreased the denitrification rate to 30 g/m^3-h.

The feasibility of membrane dissolution of hydrogen for autotrophic denitrification was demonstrated by Gantzer (1995). The study used a hollow fiber membrane system to supply high dissolved hydrogen concentration in a fixed-bed denitrification bioreactor. The hydrogen gas was maintained at a pressure of 2 atm inside the fibers. As shown in Table 1-1, the nitrate removal rate of the autotrophic process was found to be much higher than that of a heterotrophic process.

The steady state nitrate flux into the fixed bed biofilms averaged about 85 mg NO_3^--N/m^2/hr, a figure that is much higher than the denitrification rates reported in the literature. Although the author stated that the process (i.e., the biofilm) was nitrate limited, no experiment was conducted to verify this claim. Based on data reported by various researchers, it appears that the process is "liquid phase hydrogen concentration limited," which is in contrast with the results reported by Kurt et al. (1987).

1.1.9 Denitrification Costs

The costs associated with the use of methanol, ethanol, and acetic acid as denitrification substrates in a fluidized bed reactor were estimated by Croll et al. (1985). For operating conditions producing acceptable effluent nitrate concentrations, methanol was found to have the lowest cost. Acetic acid and ethanol were found to be 50% and 25% more expensive than methanol, respectively. The capital and operating cost for denitrification using an autotrophic process was found to be higher than the cost of using heterotrophic denitrification (Mateju et al., 1992).

Table 1-1. Comparison of Nitrate-N Removal Rates

Electron Donor	Nitrate Removal Rate (mg NO_3^--N/m^{2}h)	Reference and Experimental Conditions
Methanol	10	Liessens et al., 1993. Biofilm reactor, methanol
Hydrogen	13	Kurt et al., 1987. Biofilm reactor (sand), hydrogen at 1 atm
Hydrogen	33	Islam et al., 1994. Biofilm reactor, hydrogen at 1 atm
Hydrogen	85	Gantzer, 1995. Biofilm reactor, hydrogen at 2 atm
Hydrogen	58	Benedict, 1996. Biofilm reactor, hydrogen at 1 atm

However, the investigators assumed a conventional bubble aeration system for dissolving hydrogen in water. The oxygen transfer efficiency (OTE) of such systems for oxygen is about 8% to 20% (Mueller and Boyle, 1988). If the substrate is delivered in a stoichiometric quantity, the cost estimates of hydrogen and methanol change. The treatment cost for 100 lb of NO_3^--N for hydrogen, methanol, ethanol, and acetic acid has been shown in Figure 1-1 (Gantzer, 1995).

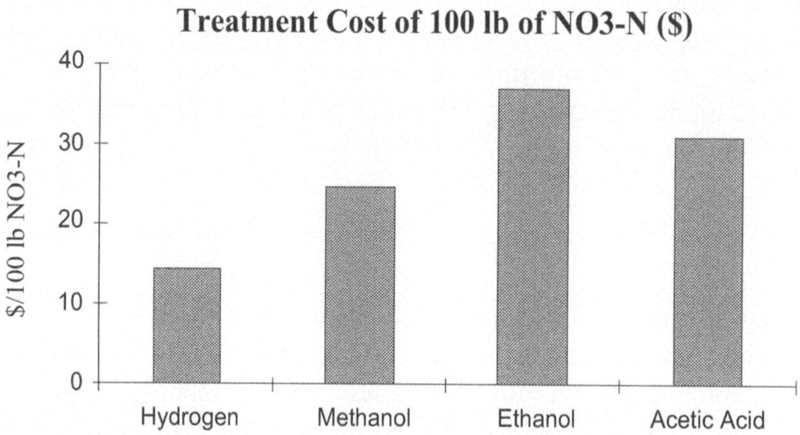

Figure 1-1. Electron Donor Costs for Biological Denitrification of 100 lb NO_3^--N

The cost analysis assumes on-site generation of hydrogen and no transportation costs for methanol, ethanol, and acetic acid. As shown in Figure 1-1, the cost of methanol is about 73% greater than the cost associated with hydrogen use. The transportation cost may be as high as the cost of the chemical itself, depending on the actual location of the site. Therefore, if the transportation costs are included in the analysis, the use of hydrogen is more economical. The costs associated with ethanol and acetic acid were calculated using the results obtained by Croll (1985).

1.2 Research Needs

To eliminate chemical contamination from added organic carbon, autotrophic, hydrogen-oxidizing bacteria can be used for denitrification. These bacteria use carbon dioxide or bicarbonate for cell synthesis. Hydrogen is used as the energy source and electron donor while nitrate acts as the electron acceptor. Autotrophic hydrogen oxidizing bacteria thus reduce nitrate to nitrogen and oxidize hydrogen gas to water as indicated in the following equation:

$$2NO_3^- + 5H_2 \rightarrow N_2 + 4H_2O + 2OH^-$$

The growth rate of autotrophic bacteria is typically slower than that of heterotrophic bacteria. Therefore, lower concentrations of biomass and soluble microbial products can be expected in the reactor effluent. The advantages of this process over conventional heterotrophic denitrification can be summarized as follows:

- No addition of organic carbon for cell synthesis;
- No interferences with the disinfection process (no disinfection byproduct formation);
- Reactor startup is simple, and autotrophic bacteria are hardy;
- The process can be operated as a fixed or fluidized bed reactor;
- Kinetic parameters of autotrophic and heterotrophic bacteria are very similar, meaning that the reactor volumes and resulting capital cost are also comparable; and

- The process is cost effective compared to usage of methanol, ethanol, or acetate as carbon sources.

The fact that hydrogen gas is inherently clean makes it the ideal reactant to denitrify drinking water. Hydrogen generates less excess biomass than the use of organic compounds and, if generated on site, is less expensive than the use of methanol, ethanol, and acetate based on bulk chemical cost.

The concerns for safety and lack of gas dissolution devices for hydrogen can be resolved by using a membrane gas transfer device. A membrane system can be used to transfer hydrogen with 100% gas transfer efficiency, and the system's bubble-free operation can prevent the release and waste of hydrogen that could lead to explosive conditions in confined spaces.

This project, therefore, addresses the feasibility of a novel membrane process for hydrogen gas transfer for autotrophic denitrification.

1.3 Project Goals and Objectives

The main goal of this study is to examine the technical feasibility of autotrophic denitrification of water and wastewater using membrane bioreactors. The specific objectives are to:

- *develop* an acclimated mixed culture capable of autotrophic denitrification using hydrogen as an electron donor;
- *evaluate* hydrogen transfer characteristics of hollow fiber membrane modules; and
- *develop* design procedures for continuous-flow, autotrophic, attached-growth, fixed-bed (plastic media) reactors.

1.4 Participants

This project was conducted as a joint effort between the Water Environment Research Foundation and Rowan University.

1.5 Organization of Report

This report includes background information and a discussion of the current status of biological denitrification processes (Chapter 1.0); results of batch autotrophic denitrification experiments (Chapter 2.0); a report on the gas transfer characteristics of hydrogen using commercial hollow fiber membranes (Chapter 3.0); design parameters for a fixed-film, autotrophic dentrification reactor with membrane gas delivery (Chapter 4.0), and a list of conclusions and recommendations (Chapter 5.0).

CHAPTER 2.0

BATCH STUDIES

2.1 Introduction

The project team carried out batch autotrophic denitrification experiments using an acclimated mixed bacterial culture obtained from a local wastewater treatment plant. Inorganic bicarbonate ions dissolved in water or carbon dioxide gas were used as the carbon source, while hydrogen gas was supplied as the electron donor. Anoxic conditions were maintained so that nitrate acted as the electron acceptor. A number of common genera of soil bacteria are able to use hydrogen as an electron donor and inorganic carbon for cell synthesis. Bacteria such as *Pseudomonas pseudoflava*, *Ralstonia eutropha* (formerly *Alcaligenes eutrophus*), and *Paracoccus denitrificans* have the ability of autotrophic denitrification using hydrogen gas as the electron donor (Schmidt et al., 1989; Selenka and Dressler, 1990).

Batch experiments were conducted in the laboratory with the following objectives:

- *develop* an acclimated mixed culture capable of autotrophic denitrification using hydrogen as an electron donor;
- *deliver* inorganic carbon through CO_2 and Na_2CO_3; and
- *monitor* nitrate, nitrite, cell mass, pH, and alkalinity concentrations with time.

2.2 Materials and Analytical Methods

Batch studies were conducted using sodium carbonate or carbon dioxide for inorganic carbon and nitrate as the electron acceptor for the autotrophic denitrification culture. Sodium carbonate (Na_2CO_3) and sodium nitrate ($NaNO_3$) were obtained from Sigma Chemicals Co., St. Louis, MO. Nutrient buffer pillows for determining biochemical oxygen demand (Hach Company, Loveland, CO) were amended with excess phosphate buffer (Ergas and Reuss, 2001) and used for the growth media. High-purity hydrogen and carbon dioxide gas were obtained from MG Gas Industries (Malvern, PA). Deionized, distilled water from a Barnstead Water Purification Water System (Dubuque, IA) was used for preparation of all solutions.

A mixed culture was obtained from the anoxic denitrification basin of the Winslow Wastewater Treatment Plant (Winslow, NJ). An acclimated culture for autotrophic denitrification using hydrogen-oxidizing bacteria was developed in batch reactors at room temperature.

The following parameters were monitored during the course of the batch experiments: pH, alkalinity, nitrate, nitrite, and cell mass. Nitrate and nitrite were measured using standard method 4500-B (AWWA et al., 1995) and Hach method 8153, on a Hach DR 4000U (Loveland, CO) spectrophotometer. pH was measured with an Orion 720 pH Meter (Beverly, MA). Alkalinity was determined using a Hach digital titrator according to standard method 2130 B (AWWA, 1995). Dissolved oxygen concentrations were measured with a YSI 520 DO Meter (Yellow Springs, CO). Cell mass was measured as protein according to the method of Lowry et al. (1951) and as mixed liquor suspended solids (MLSS) using a standard method (AWWA, 1995). Bovine serum albumin (Sigma Chemicals, St. Louis, MO) was used as the protein standard. Dissolved hydrogen concentration was measured with an Orbisphere Micro Logger model 3654 (Orbisphere Laboratories, Neuchatel, Switzerland).

2.3 Experimental Setup

The experimental setup for the batch studies is presented in Figure 2-1. Amber glass jars (4 L) equipped with magnetic stirring bars for adequate mixing were used as the reactors. Two sets of batch studies were conducted with CO_2 gas and Na_2CO_3 for the delivery of inorganic carbon.

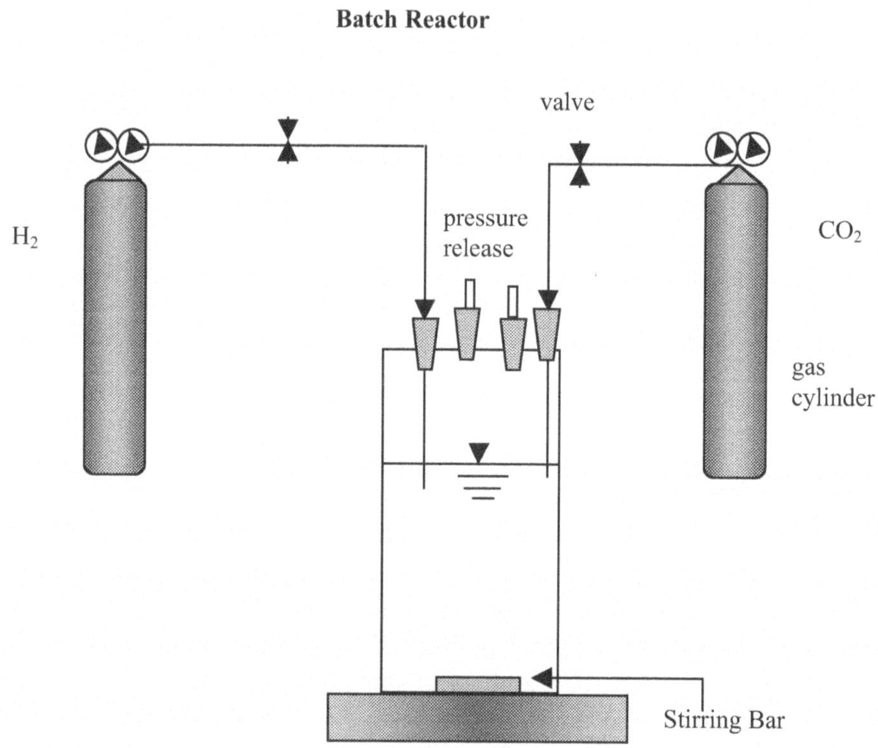

Figure 2-1. Experimental Setup for Batch Studies

Control reactors were maintained in both sets of experiments. The hydrogen concentration was not measured in all batch reactors. One reactor was dedicated for hydrogen concentration measurements for each set.

2.3.1 Batch Studies with Na_2CO_3

Four batch reactors were used in this study. Each batch reactor received 100 mg/L of Na_2CO_3 and 100 mg/L of NO_3^-. Therefore, the reactors had a concentration of 25 mg/L NO_3^--N. The reactors were seeded with 4 mL of mixed liquor from an acclimation reactor containing bacteria acclimated to $CO_3^=$ as the sole carbon source. The control reactor did not receive any mixed liquor. The reactors were purged with H_2 gas every 4 hours with the aid of solenoid valves (Burkert, Orange, CA), which are controlled by programmable timers (Model XT, ChronTrol Corp., San Diego, CA). The gas lines were fitted with check valves to prevent back flow of gases. A stirring bar continuously mixed the reactor contents. Samples were analyzed for nitrate, nitrite, cell mass, hydrogen, pH, and alkalinity with time. Experiments were conducted under a fume hood.

2.3.2 Batch Studies with CO_2

Four batch reactors were also used in this study. A CO_2-based acclimation reactor was started using seed from the Na_2CO_3 acclimation reactor. The reactors were purged every 4 hours with H_2 gas and every 10 hours with CO_2 gas through a bubble stone. The gas cycling was controlled with the aid of solenoid valves (Burkert, Orange, CA) and a programmable timer (Model XT, ChronTrol Corp., San Diego, CA). Samples were analyzed for nitrate, nitrite, cell mass, hydrogen, pH, and alkalinity with time. Experiments were conducted under a fume hood.

2.4 Results

Results for nitrate removal from batch studies conducted with sodium carbonate are presented in Figure 2-2. Nitrate removal was almost complete in 2 days. Nitrite concentration indicates some accumulation but complete removal during the duration of the experiment. Other researchers (Kurt, Dunn, and Bourne, 1987; Oh et al., 1999; Miyake et al., 1996; Smith et al., 1993) have reported nitrite accumulations during batch autotrophic denitrification studies. All authors suggested that a continuous flow process having a relatively long retention time would result in complete nitrogen removal. Cell mass, pH, and alkalinity concentrations for these experiments are presented in Figures 2-3, 2-4, and 2-5. The buffering capacity in the reactors seems adequate from these results.

Results from the batch experiments with CO_2 were very similar to those obtained from the sodium carbonate tests. Results for nitrate, nitrite, cell mass, pH, and alkalinity are presented in Figures 2-6 through 2-9. In both systems the total carbonate distribution between H_2CO_3 and HCO_3^- would have to be similar because the system pH were similar. This would explain the similar nitrate and nitrite removal in both sets of experiments. The only difference noted was in cell mass measurement techniques. Protein measurements worked well with the CO_2 system, and suspended solids measurements worked for the Na_2CO_3 system. Cell mass as protein was analyzed according to the method of Lowry et al. (1951). Protein concentrations were analyzed because problems were encountered for representative cell mass concentration as MLSS.

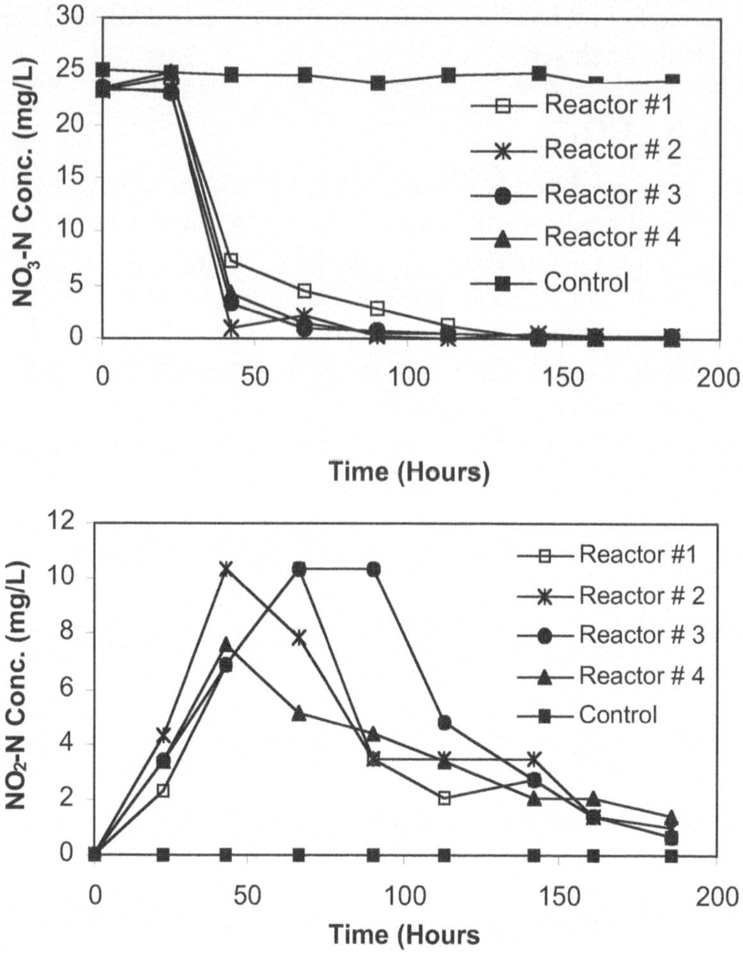

Figure 2-2. Nitrate and Nitrite Concentration with Time in Batch Reactors with Na_2CO_3

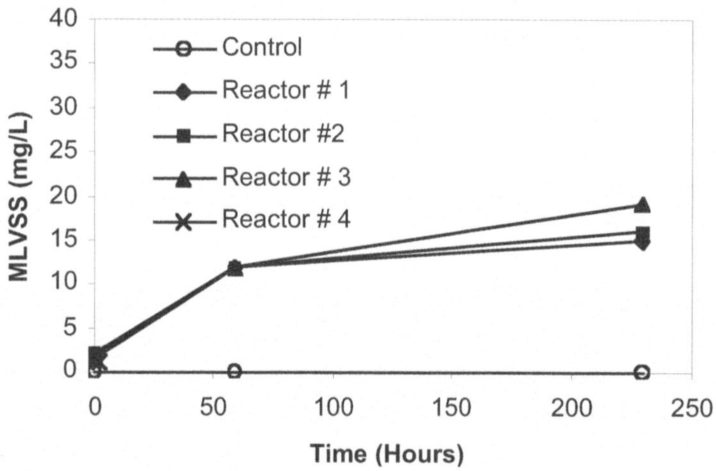

Figure 2-3. Cell Mass Concentration with Time in Batch Reactors with Na_2CO_3

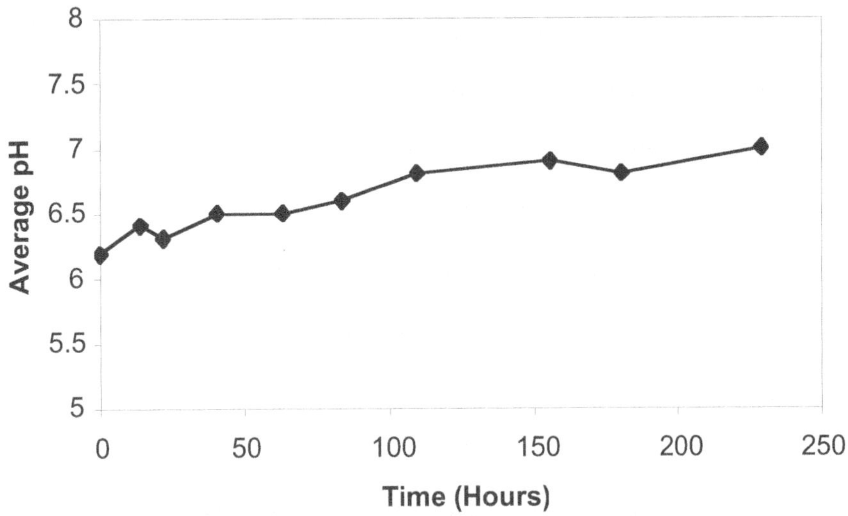

Figure 2-4. pH Changes with Time in Batch Reactors with Na_2CO_3

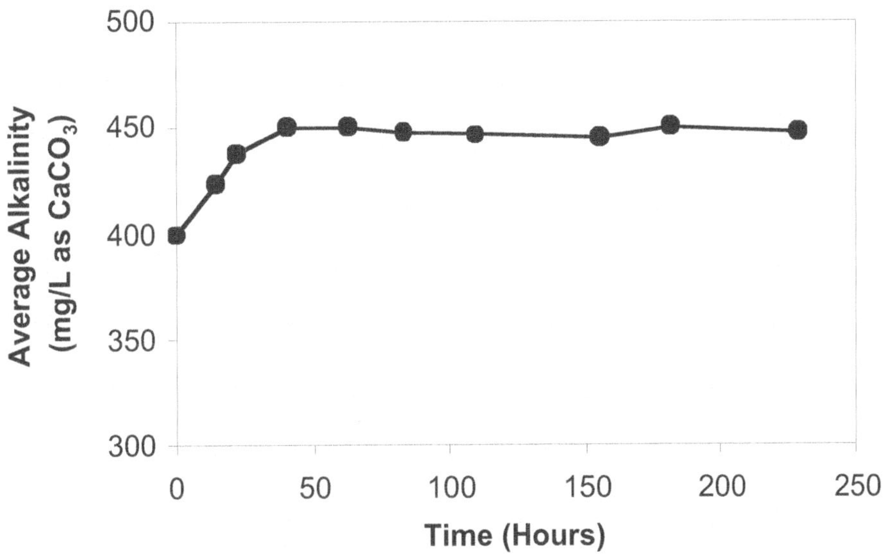

Figure 2-5. Alkalinity Changes with Time in Batch Reactors with Na_2CO_3

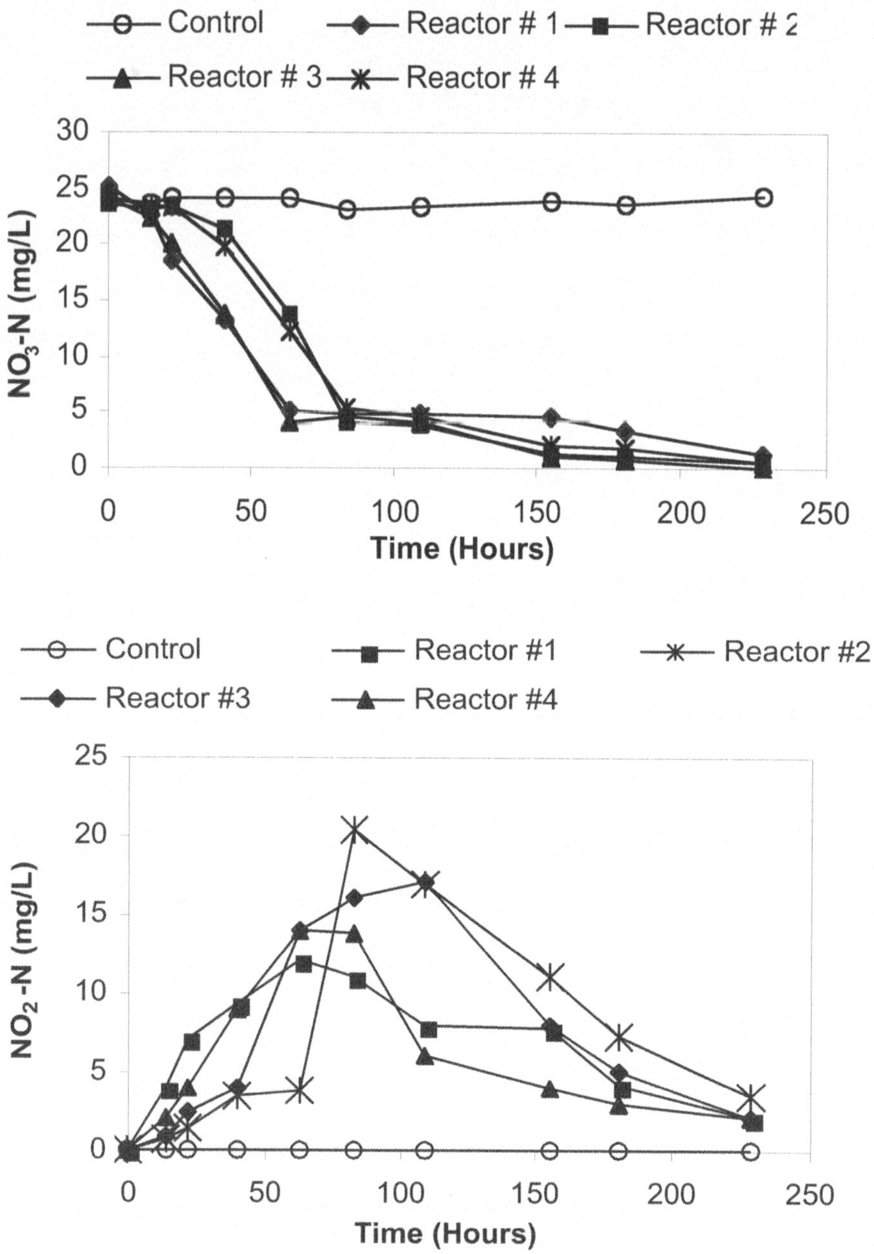

Figure 2-6. Nitrate and Nitrite Concentration with Time in Batch Reactors with CO_2

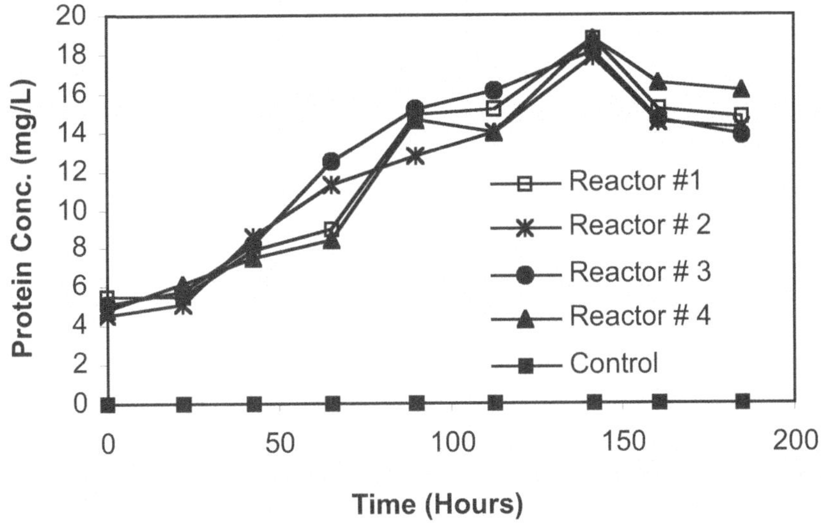

Figure 2-7. Cell Mass Concentration with Time in Batch Reactors with CO_2

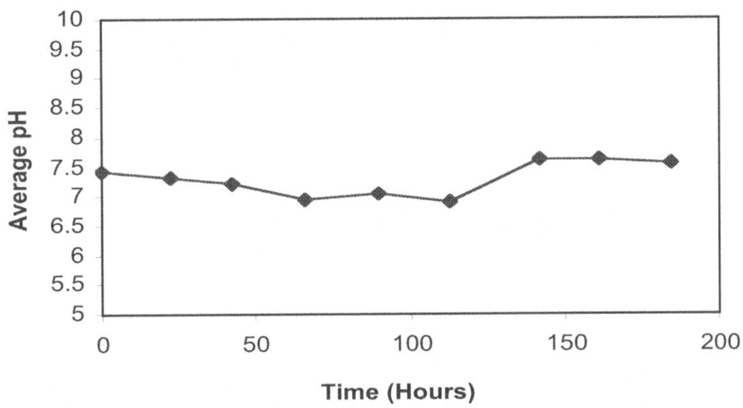

Figure 2-8. pH Changes with Time in Batch Reactors with CO_2

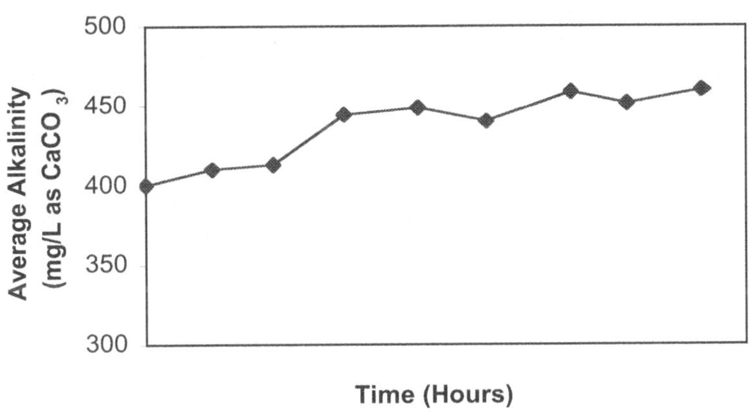

Figure 2-9. Alkalinity Changes with Time in Batch Reactors with CO_2

Hydrogen concentrations were measured intermittently during the course of these experiments. Hydrogen concentration was found to be 62.5% of saturation in both systems. This indicates that the process was not hydrogen limiting. All batch studies indicated that rapid nitrate removal is possible with autotrophic denitrifiers. Nitrite accumulations occurred but were removed during the course of these experiments.

2.5 Conclusions

These studies indicate the existence of hydrogen oxidizing denitrifiers in wastewater treatment plants. A mixed culture capable of utilizing an inorganic carbon source could be developed having the ability of using hydrogen as an electron donor. Adequate pH control was possible, and the pH averaged around 6.6 and 7.3 for the two sets of experiments. Hydrogen concentration was found to be 62.5% of saturation in both systems, indicating that the process was not hydrogen limiting. All batch studies indicate that rapid nitrate removal is possible with autotrophic denitrifiers.

CHAPTER 3.0

Hydrogen Transfer Via Hollow Fiber Membranes: Experimental Studies

3.1 Introduction

Hydrogen transfer into water is needed in water and wastewater treatment where it is an energy source for certain biological processes, such as autotrophic denitrification. In recent years hydrogen transfer has also become important in natural attenuation (EPA, 1987) and in-situ biodegradation of organic pollutants (Novak et al., 2001). Compared to other, conventional organic energy sources, hydrogen is less expensive and does not leave an organic residue in the treated water. Nevertheless, the use of hydrogen has been limited because of the safety issue associated with conventional sparging of hydrogen into water: formation of bubbles presents an explosion hazard. Hollow fiber membranes may be used for bubbleless hydrogen-to-water transfer.

3.1.1 Membrane Selection

Hydrogen gas transfer into water may be accomplished through porous, nonporous, and asymmetric membranes. To obtain a high hydrogen transfer rates, the membrane material must be (1) able to withstand high pressures; (2) highly hydrogen permeable; and (3) capable of being formed into small-diameter hollow fibers.

Microporous hydrophobic membranes have the advantage of very high gas permeabilities. Because the pores are gas-filled, diffusion takes place in the gas phase at a much faster rate than through a polymer film. These membranes are available as small-diameter (200–400 μm) hollow fibers that provide a very high specific surface area (i.e., surface area to volume ratio). The use of porous membranes is limited to low pressure applications because high pressures result in bubble formation. Nevertheless, they have found application in absorption and water treatment. Hydrogen mass transfer was achieved through polytetrafluoroethylene (PTFE) and Teflon hollow-fiber membranes as a valid alternative to absorption tower operation (Iversen et al., 1995). Microporous hollow fiber membranes were applied in hydrogen gas scrubbing and absorption (Li et al., 2000). More recently, hydrogen was transferred through hollow fiber membranes directly to a bacterial culture for biological denitrification of drinking water (Ergas and Reuss, 2001).

Nonporous membranes, such as silicone rubber, may be operated at elevated gas pressures without bubble formation. Gas transfer is accomplished via a solution-diffusion mechanism. Hydrogen first dissolves in the membrane material and subsequently diffuses through the membrane to the liquid. Brautigam and Sekoulov (1987) introduced hydrogen to water through silicone rubber tubular membranes for the purpose of biologically treating wastewater through denitrification in a fluidized bed with hydrogen as the electron donor. There are several disadvantages of non-porous film in hydrogen transfer. Since non-porous membranes are thicker than porous membranes, they present an increased resistance to mass transfer, and, therefore, reduce hydrogen transfer rate. In addition, non-porous membranes are not available in small diameter (100-400 μm) fiber form, which decreases the area available for mass transfer.

Anisotropic or asymmetric membranes consist of a thin skin membrane layer cast on top of a thicker porous backing. The thin skin performs the separation, while the thick layer provides mechanical support for high-pressure applications. Sell et al. (1993) applied a non-porous polymer composite membrane in hollow fiber configuration for hydrogen transfer. The outer surface of the fibers was composed of a polyetherimide nonporous layer and a microporous skin on the lumen side. The internal skin was coated with polydimethylsiloxane (PDMS), resulting in 1.2 mm outer diameter and 1 mm inner diameter of the hollow tube. Hydrogen gas was transferred to a bacterial biofilm through an asymmetric composite membrane (Lee and Rittmann, 2000). The outer and inner layers were composed of microporous polyethylene with a nonporous polyurethane layer sandwiched between them. As is the case of nonporous membranes, an anisotropic film is comparatively thick, thereby limiting its hydrogen transfer characteristics and forcing an increase in the diameter of hollow fibers. And last, asymmetric hollow fiber membranes are much more expensive than microporous membranes because of an involved spinning and coating manufacturing process.

3.2 Membrane Module Design

Gas transfer via hollow fiber membrane modules may be accomplished in flow-through mode or in a sealed-end configuration as shown in Figure 3-1. In the flow-through mode (Figure 3-1a), gas is pumped continuously through the lumen of the fiber. A portion of the gas diffuses through the membrane, and the remaining gas is vented off (Yang and Cussler, 1986). Because gas is vented off in the flow-through mode, it is impossible to achieve 100% gas transfer efficiency. Also, flow-through operation may result in effective stripping of dissolved volatile compounds from the liquid with the venting gas.

In the sealed-end operational configuration (Figure 3-1b), one end of each hollow fiber membrane in the bundle is sealed. Membranes are then pressurized with gas, which diffuses through the membrane into the water. Ahmed and Semmens (1992) demonstrated that sealed-end operation results in superior performance as compared to the flow-through configuration.

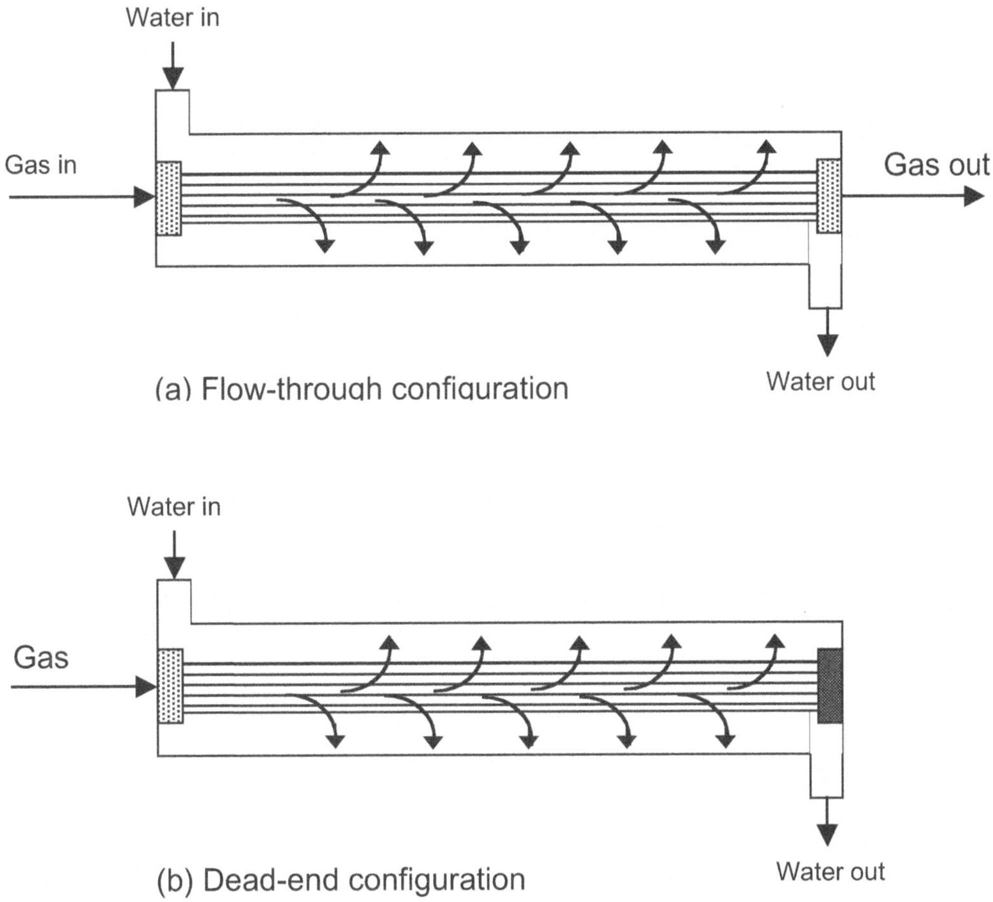

Figure 3-1. Flow-Through and Sealed-End Modules

Both of those considerations are important to hydrogen gas transfer. Venting hydrogen, which results from flow-through operation, constitutes a safety hazard. Additional costs must be incurred to contain the offgas and to avoid explosion risks when mixed with air. Stripping of compounds such as volatile organic compounds (VOCs) presents a problem in applications of hydrogen transfer to wastewater treatment. Moreover, VOCs present in wastewater may diffuse through the membrane into the hydrogen stream, increasing the cost of subsequent treatment of the offgas. These constraints may be effectively eliminated through sealed-end operation. All hydrogen diffuses into the water through the membrane, resulting in 100% transfer efficiency. Elimination of the vent stream reduces the explosion risk hazard and prevents stripping of VOCs.

Therefore, the use of microporous hollow fiber membranes for hydrogen gas transfer into water was evaluated. Mass transfer correlations for hydrogen were developed in two configurations: sealed-end and flow-through. Experiments were conducted with the following objectives:

- *Construct* hollow fiber membrane modules for hydrogen gas delivery;
- *Evaluate* the effect of various physical and process variables on the mass transfer rate across the membrane system; and
- *Determine* the mass transfer coefficient correlations for various operating parameters such as flow and pressure.

3.3 Experimental Methods

A schematic of a flow-through module is shown in Figure 3-2. Hollow fiber membranes are potted into a tube to form a shell-and-tube configuration. The potted end is connected to an air pump. Gas transfer occurs as water is pumped co-currently on the outside of the fibers on the shell side. The turbulence of water flow causes the fibers to fluidize, or flutter independently. This fiber movement within the pipe enhances mass transfer and may minimize biological and chemical fouling on the outside of the fibers.

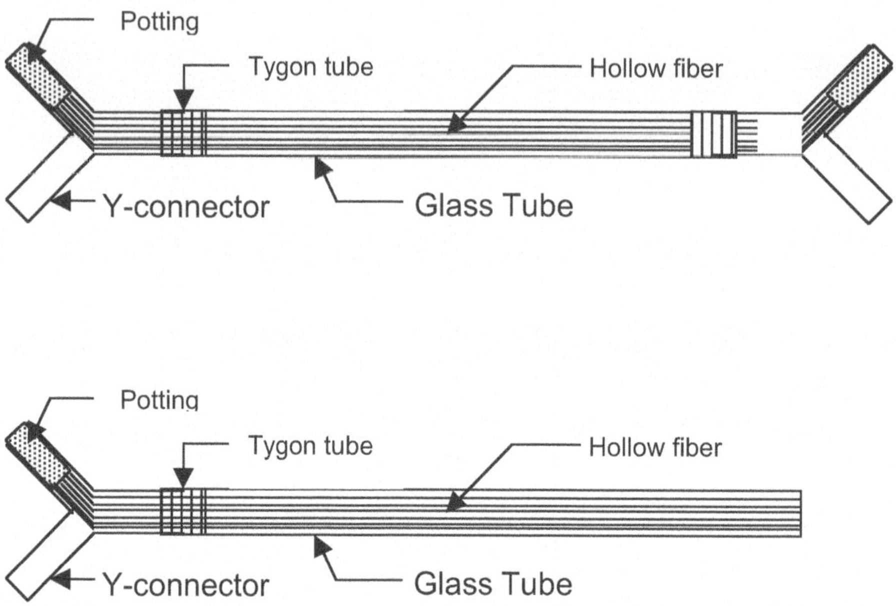

Figure 3-2. Schematic Diagram of Membrane Modules

Celgard X40-200 microporous hollow fiber membranes (Celgard Inc., Charlotte, NC) were selected for this study. These are thin-walled, opaque, polypropylene membranes with a nominal internal diameter of 200 μm. Nominal porosity is 25%, effective pore size of 0.04 μm, pore dimensions of 0.04 μm × 0.01 μm. They are also durable and pressure resistant with burst strength of 400 psi and tensile break strength exceeding 300 g/fiber.

The hollow fiber module was constructed by potting several fibers into an external shell. The shell consisted of a glass tube fitted with a Y-connector. After one end of each hollow fiber was heat sealed, the open ends were potted into the Y-connector with an epoxy adhesive (DP-125, 3M, St. Paul, MN) as shown in Figure 3-2. The physical characteristics of the membrane modules tested are summarized in Table 3-1.

Table 3-1. Dimensions of Hollow Fiber Membrane Modules

No. of Fibers	Fiber Diameter (cm)	Length (cm)	Equivalent Diameter (cm)	Shell Diameter (cm)
36	0.03	125	0.195	0.6
6	0.03	69	0.454	0.6
1	0.03	80	0.57	0.6
1	0.03	71	0.37	0.4
6	0.03	63.5	0.266	0.4

The experimental setup is shown in Figure 3-3. The water was pumped from a 2.5-L closed reservoir to the hollow fiber membrane module by a peristaltic pump (Model 7553-70, Cole-Parmer, Vernon Hills, IL). The water flow rate was monitored with a turbine flow meter (Model 32919-25, Cole-Parmer, Vernon Hills, IL). High purity hydrogen, obtained from MG Gas Industries (Malvern, PA), was supplied from a gas cylinder and the pressure was measured with a pressure transducer (Model PX 105, Omega Engineering, Stamford, CT). Mixing in the reservoir was induced with a magnetic stirrer. The reservoir was sealed to avoid gas exchange with the atmosphere, and a check valve was used on the reservoir vent. The membrane module was oriented horizontally.

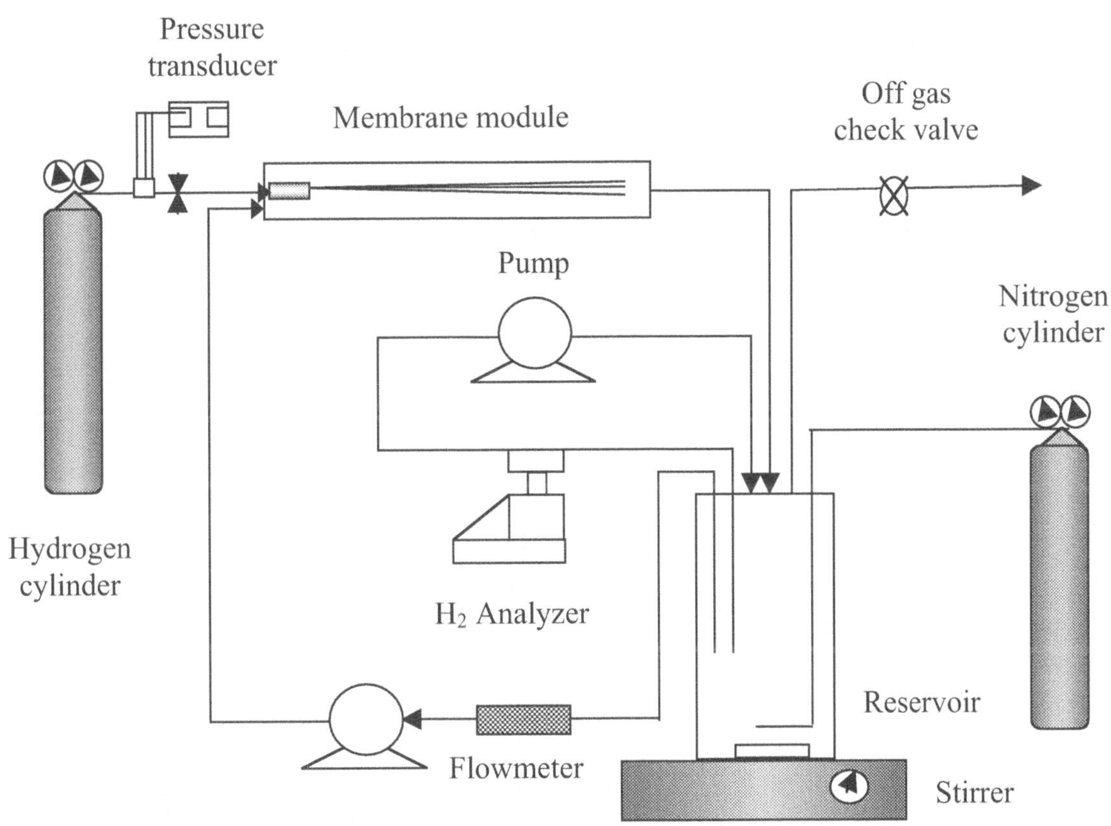

Figure 3-3. Hydrogen Transfer Experimental Setup

The static, or gassing out, method was used to measure the hydrogen transfer coefficient in the hollow fiber membrane module. Hydrogen transfer occurred as water was pumped through the hollow fiber membrane module resulting in an increase in dissolved hydrogen concentration with time. Dissolved hydrogen concentration was measured with an Orbisphere Micro Logger (Model 3654, Orbisphere Laboratories, Neuchatel, Switzerland) in 15-second intervals for the duration of 0.5 hours. Water temperature and atmospheric pressure were recorded for each test.

3.4 Data Analysis

In a microporous hollow fiber module, the driving force for hydrogen transfer is provided by the hydrogen concentration gradient between the gas and the liquid phase across the membrane. The hydrogen transfer involves three sequential steps. First, the hydrogen molecules diffuse out of the bulk air phase to the membrane surface. Second, they diffuse through the gas filled pores in the walls of the hydrophobic hollow fibers. Finally, when the hydrogen molecules reach the other side of the membrane, they diffuse into the water phase. The mass transfer coefficient (K), or resistance to mass transfer ($1/K$), is given by

$$\frac{1}{K} = \frac{1}{k_L} + \frac{1}{k_M H} + \frac{1}{k_G H} \qquad (3.4\text{-}1)$$

where k_L, k_M, and k_G are the individual mass transfer coefficients in the liquid, across the gas filled membrane pores and in the gas phase, respectively; H is the Henry's law constant. Gas diffusion coefficients are approximately four orders of magnitude faster than diffusion coefficients in water. Because both k_G and k_m are characterized by gas phase diffusivities, they are expected to be much larger in value than k_L. In addition hydrogen is only slightly soluble in water (i.e., the value of H is large), the resistances $\frac{1}{Hk_G}$ and $\frac{1}{Hk_M}$ are also expected to be negligible compared to $\frac{1}{K_L}$. Therefore, hydrogen transfer is expected to be limited by liquid film diffusion in the membrane system.

The effects of various physical and process variables on the mass transfer rate across the membrane system were evaluated. Physical parameters manipulated in this study were (1) shell diameter; and (2) length of fiber. Increasing the pumping rate increased the water velocity outside the fibers.

A sealed-end hollow fiber contactor was constructed by potting the ends of a desired number of hollow fibers into a shell, usually a glass or a PVC pipe. An impulse heat sealer was used to seal the other end of the fibers. The gas phase flows inside the fibers, whereas the liquid phase flows outside the fibers. The two streams were co-current as shown in Figure 3-3. Because the membrane lumen is constantly replenished by fresh air, the equation for rate of gas transfer can be written as:

$$N_l = - K\, A\, C_L \qquad (3.4\text{-}2)$$

where N_1 is the mass of hydrogen transferred per unit time, K is the overall mass transfer coefficient, A is the interfacial area of transfer, and C_L is the concentration of hydrogen in the liquid.

Equation 3.4-2 is usually written in concentration units by dividing by the volume V of the system:

$$\frac{N_l}{V} = \frac{dC_L}{dt} = -K\frac{A}{V}C_L = -K\,a\,C_L \qquad (3.4\text{-}3)$$

where $a = \frac{A}{V}$ = interfacial area per unit volume. In a membrane system, the value of a, the specific surface area, can be calculated from the surface area of the fibers. Then Equation 3.4-3 can be written as

$$\frac{dC_L}{dt} = -K\frac{n\pi dL}{V}C_L$$

or, $$v_L\frac{dC_L}{dz} = -Ka\,C_L$$

or, $$dC_L/C_L = -(Ka/v_L)\,dz$$

where L, d and n are the length, diameter, and number of the fiber respectively. Integration of the equation from C_1 (inlet hydrogen concentration) to C_2 (exit hydrogen concentration) gives

$$C_2 = C_1\exp(-Ka\frac{L}{v_L}) \qquad (3.4\text{-}4)$$

where L is the length of the fibers and v_L is the velocity of liquid flow outside the fiber. This equation is used to analyzc the experimental data obtained from the tests.

The water from the module recirculated back to the reservoir where the hydrogen concentration change was measured with time. The mass balance for the reservoir gives:

$$V\frac{dC_1}{dt} = Q\,C_2 - Q\,C_1 \qquad (3.4\text{-}5)$$

where V is the volume of the reservoir, and Q is the flowrate of water. Substituting the values of C_2 from Equation 3.4-4 into Equation 3.4-5 and simplifying the expression, gives:

$$V\frac{dC_1}{dt} = Q\left\{C_1\exp(-Ka\frac{L}{v_L}) - 1\right\}$$

This equation is subject to the following boundary conditions:

$t = 0, \quad C_l = C_0$ (initial reservoir hydrogen concentration)
$t = t, \quad C_l = C$

Integrating, we obtain the following exponential equation:

$$\ln\left(\frac{C}{C_0}\right) = \frac{Q}{V}\left[\exp\left(-\frac{KaL}{v_L}\right) - 1\right]t \qquad (3.4\text{-}6)$$

Therefore, the overall mass transfer coefficient can be determined from the slope of ln (C/C_0) time plot.

There are two important assumptions in the above development:

1. The experiment was always conducted in such a way that the residence time in the reservoir was large compared to the response time of hydrogen probe; and
2. The concentration of hydrogen in the reservoir changes slowly compared to the concentration changes occurring in the module for the assumption of steady state in the module to be realistic.

The effect of water flow rate on the hydrogen mass transfer coefficient was investigated. Experiments were completed with various pumping speeds in the range of 0.2 L/min to 0.6 L/min while hydrogen gas pressure on the inside of hollow fibers was held constant.

3.5 Results and Discussion

Laboratory-scale experiments were conducted to evaluate the influence of operating conditions on the performance of the module. In each of these tests, the reservoir concentration data generated a linear relationship when plotted in the form of Equation 3.4-6:

$$\ln\left(\frac{C}{C_0}\right) = \frac{Q}{V}\left[\exp\left(-\frac{KaL}{v_L}\right) - 1\right]t$$

A typical plot of ln (C/C_o) versus time is illustrated in Figure 3-4. The overall mass transfer coefficient can be extracted from the slope of the line using experimental operating conditions and module parameters.

To normalize the data, it is convenient to express these data in terms of the dimensionless Sherwood Number (*Sh*), Reynolds Number (*Re*), and Schmidt Number (*Sc*) as follows:

$$Sh = \alpha\, Re^{\beta}\, Sc^{\gamma} \qquad (3.5\text{-}1)$$

where α, β, and γ are constants. The definitions of *Sh, Rd*, and *Sc* are presented in Table 3-2. The Schmidt number is defined as the ratio of the kinematic viscosity of a fluid to the diffusivity of the solute in that fluid. Because hydrogen was the only constituent considered in this study and test temperature only varied within a few degrees, the Schmidt number may considered to be constant, and a 1/3 power dependence was assumed from heat transfer literature (Bennett and Myers, 1982).

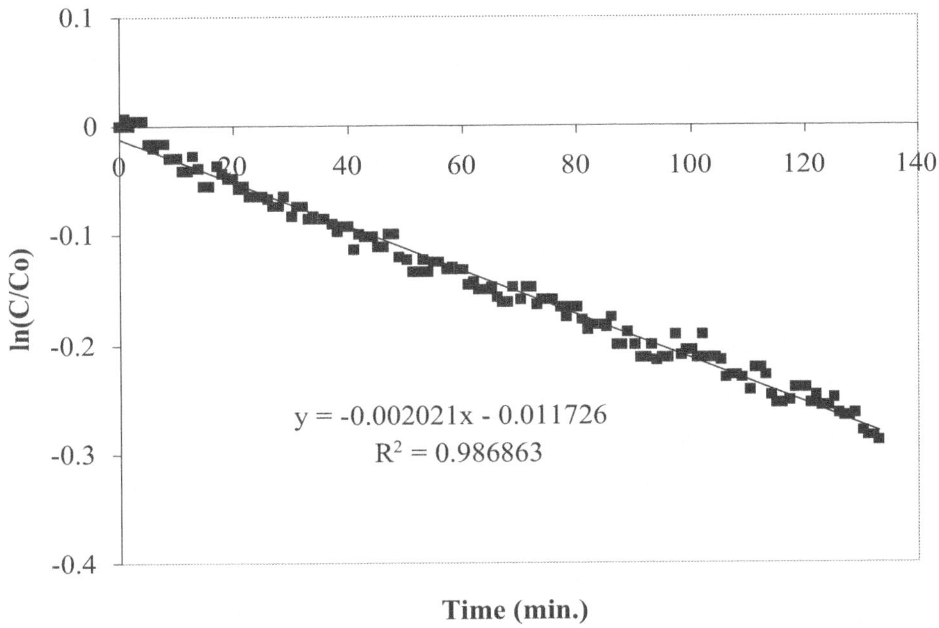

Figure 3-4. Sample ln (C/Co) Versus Time Plot

Table 3-2. Definitions of the Dimensionless Numbers in Mass Transfer

Mass Transfer		
Name	**Symbol**	**Definition**
Reynolds Number	*Re*	$\frac{dv}{\nu}$
Schmidt Number	*Sc*	$\frac{\nu}{D}$
Sherwood Number	*Sh*	$\frac{dk_L}{D}$

When fluid flow in conduits (in this case hollow fibers) is considered, it is also necessary to determine a "characteristic length" term for Reynolds number calculations. The characteristic length is defined as the effective diameter, d_e (Yang and Cussler, 1986):

$$d_e = \frac{4(\text{cross-sectional area of flow})}{\text{wetted perimeter}} \qquad (3.5\text{-}2)$$

Assuming that the mass transfer process is controlled by the liquid phase, the Sherwood number and the Reynolds number were calculated using equivalent diameter of the module, d_e and the gas phase characteristics (i.e., liquid velocity, v_L, kinematic viscosity of water, ν, and molecular diffusion coefficient of hydrogen in water at infinite dilution, D_{ab}). The experimental data were plotted as Sherwood number versus Reynolds number in Figure 3-5.

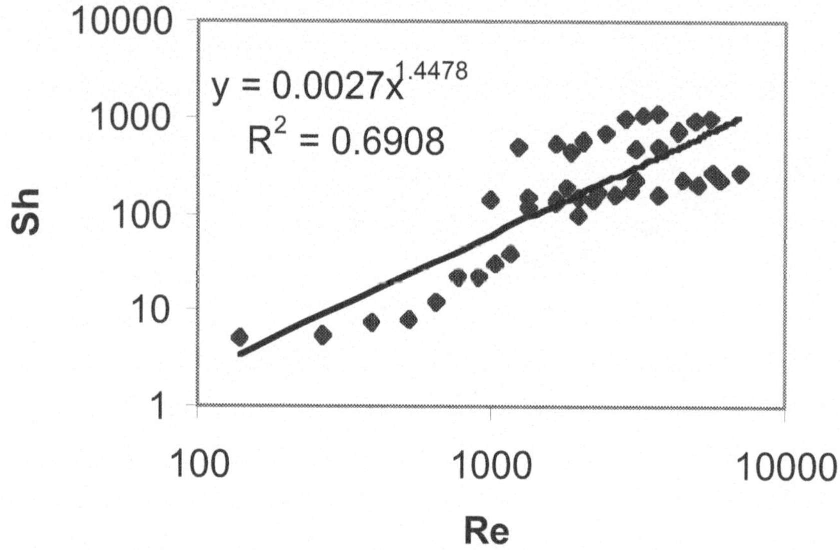

Figure 3-5. Mass Transfer Correlation for Designing Hollow Fiber Membrane Modules (*Sh* vs. *Re*)

A least squares regression analysis of the data yielded the following correlation:

$Sh = 0.0027\ Re^{1.44}$

or, $Sh = 0.00043\ Re^{1.44}\ Sc^{0.33}$ (3.5-3)

The plot shown in Figure 3-5 shows a significant amount of scatter ($R^2 = 0.69$); however, the relationship is significant at the 95% confidence level.

Yang and Cussler (1986) found that the use of the ratio d_e/l to modify Reynolds number improved the correlation with the Sherwood number. The experimental data thus were plotted as Sherwood number versus *Re*de/L in Figure 3-6 to obtain the following correlation:

$Sh = 16.77\ Re\,de/L^{1.02}$

or, $Sh = 2.68\ Re\,de/L^{1.02}\ Sc^{0.33}$ (3.5-4)

The modified correlation was successful in reducing the scatter ($R^2 = 0.83$) in the hydrogen transfer data collected in this study.

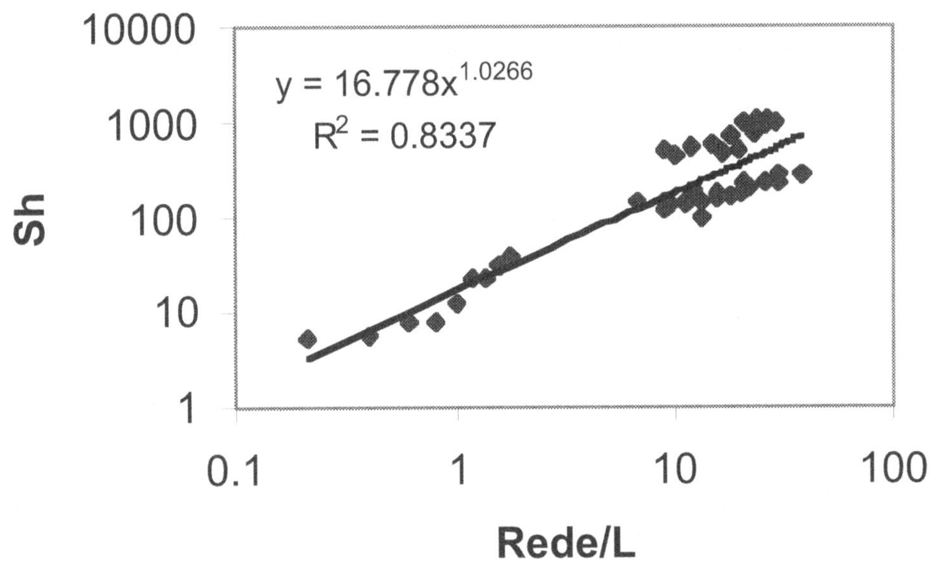

Figure 3-6. Mass Transfer Correlation for Designing Hollow Fiber Membrane Modules (*Sh* vs. *Rede*/L)

In Table 3-3 the mass transfer correlation developed in this work is compared with those found in the literature. The correlation indicated that the results of this study are very similar to those expected for flow in small tubes. The exponent of *Re* for hydrogen transfer was found to be large compared to those reported for oxygen transfer experiments. The coefficient in front of the present mass transfer correlation is also small—about two orders of magnitude smaller than the values reported by other researchers (Yang and Cussler 1986; Ahmed and Semmens, 1992) for similar hollow fiber membrane systems.

Table 3-3. Comparison of Correlations from Heat and Mass Transfer Literature

Reference	Correlation	*Re* Range
Knudsen and Katz (1958)	$Sh = 0.022\ Re^{0.6}\ Sc^{0.33}$	-
Yang and Cussler (1986)	$Sh = 1.25\ (Red_e/L)^{0.93}\ Sc^{0.33}$	5–3000
Cote et al. (1989)	$Sh = 0.61\ Re^{0.363}\ Sc^{0.33}$	0.6–49
Ahmed and Semmens (1992a,b)	$Sh = 0.108\ Re^{0.81}\ Sc^{0.33}$	300–46,000
	$Sh = 0.886\ (Rede/L)^{0.91}\ Sc^{0.33}$	
This Study	$Sh = 0.00043\ Re^{1.44}\ Sc^{0.33}$	350–4200
	$Sh = 2.68\ (Rede/L)^{1.02}\ Sc^{0.33}$	

3.6 Conclusions

A membrane process for transferring hydrogen gas to water was studied using microporous membranes. The membrane is hollow fiber microporous polypropylene and is hydrophobic to prevent wetting of membrane pores by water. The hydrogen transfer is primarily controlled by liquid film diffusion. The following mass transfer correlation can be used to design membrane modules for hydrogen transfer:

$$Sh = 2.68\ Re\mathrm{de/L}^{1.02}\ Sc^{0.33}$$

The membrane hydrogen transfer appears to be a viable technology for transferring hydrogen gas to water. Further investigation is required to evaluate large-scale applications and other operating concerns.

CHAPTER 4.0

CONTINUOUS-FLOW, FIXED-FILM REACTOR STUDIES

4.1 Introduction

One of the main objectives of this study was to evaluate the performance of denitrifying autotrophic hydrogen-oxidizing bacteria in bench-scale reactors. To evaluate the feasibility of supplying hydrogen via a membrane module, an upflow fixed-film reactor was constructed and operated in the laboratory. The reactor was inoculated with hydrogen-oxidizing bacteria and operated in batch mode until a stable biofilm developed on the media. In continuous flow mode, an anoxic inorganic nutrient buffer solution containing 22.6 mg NO_3^--N/L was fed to the reactors at varying hydraulic retention times.

4.2 Fixed-Bed Bioreactor

The fixed-bed reactor consisted of a cylindrical clear acrylic column, 2.54 cm in diameter and 14 cm in height. To ensure watertight seals, the top and the bottom of the reactor were fitted with rubber stoppers. Each rubber stopper had two openings for pumping water in and out of the reactor. One inlet and outlet set was used to supply and withdraw synthetic-nitrate-contaminated water to the system; the other set was used for recycling the liquid through the hollow fiber membrane module for hydrogen addition. The schematic of the reactor components are shown in Figure 4-1. Polymeric rings were used as fixed media in the reactor. The dimensions of the rings are shown in Figure 4-2. Although the specific gravity of the rings was less than 1, the media was tightly packed and there was no empty space for fluidization. The reactor thus operated in a fixed film mode. The effective surface area available in the reactor for biofilm growth and attachment was approximately 0.0697 m^2. The media depth and volume was constant at all hydraulic retention times evaluated.

During startup, the reactor was operated in a batch mode to develop an acclimated, hydrogen-oxidizing biofilm on the fixed media. The column was connected to a stirred tank that contained 2 L of a degassed, inorganic nutrient buffer solution containing 100 mg/L nitrate and 100 mg/L sodium carbonate to which 83.3 mL of acclimated culture seed was added. The degassed buffer solution was saturated with hydrogen gas at the beginning, and then the gas was sparged into the stirred tank in 5-sec pulses at 2-h intervals using a solenoid valve (Burkert, Inc., Orange, CA) and a timer (Model XT, ChronTrol Corp., San Diego, CA). The reactor content was withdrawn from the stirred tank and recycled through the column using a peristaltic pump (Model 7553–70(20), Cole-Parmer, Niles, IL) at a flow rate of 220 mL/min. The pH was maintained at 7.0 with carbon dioxide gas addition periodically. The schematic of the apparatus used for start-up is provided in Figure 4-3.

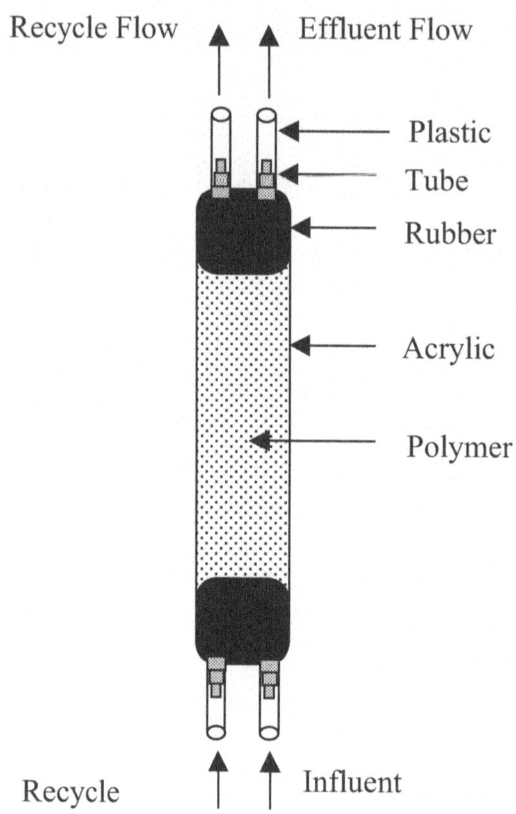

Figure 4-1. Schematic of the Fixed Biofilm Reactor

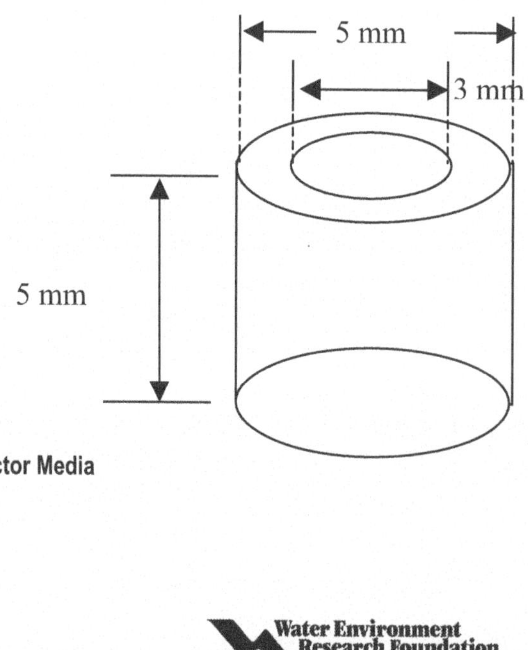

Figure 4-2. Polymeric Reactor Media

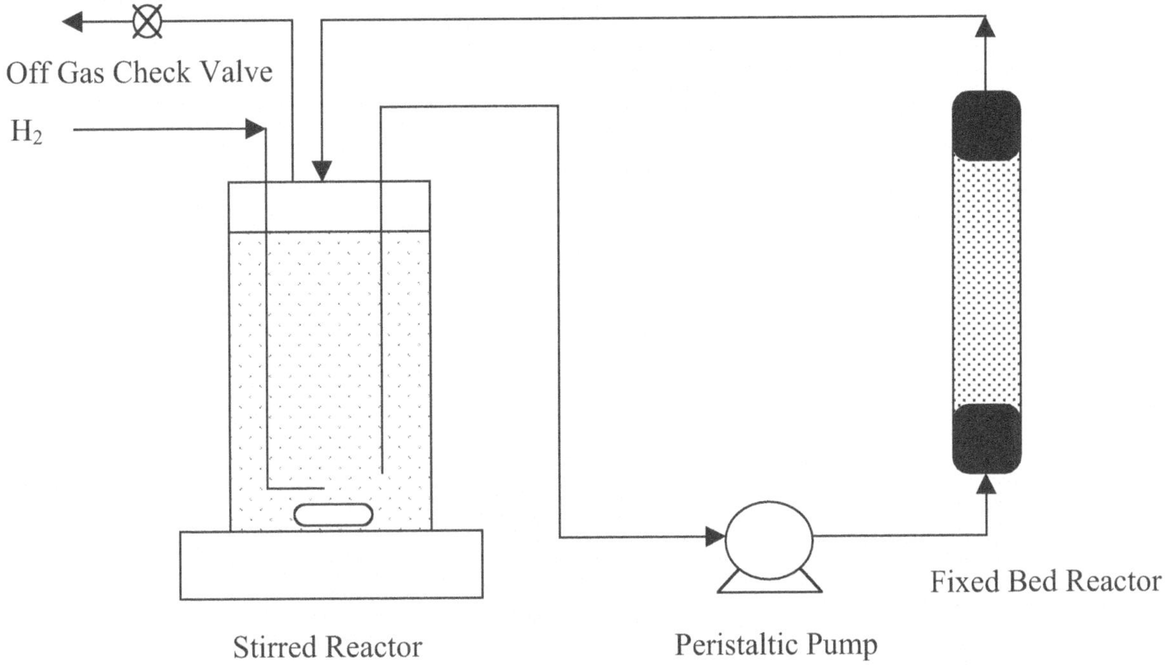

Figure 4-3. Fixed Biofilm Reactor Startup Configuration

4.3 Hollow Fiber Membrane Module

A schematic of the sealed-end hollow fiber module was presented in Figure 3-2. Membrane specifications and membrane module construction details were presented in Section 3.2 of Chapter 3.0.

4.4 Materials and Analytical Methods

Continuous-flow studies were conducted using carbonate as sole carbon source and nitrate as the electron acceptor for the autotrophic denitrification culture. Sodium carbonate (Na_2CO_3), sodium nitrate ($NaNO_3$), phosphate buffer (KH_2PO_4 and K_2HPO_4) and micronutrients ($MgSO_4$ $7H_20$, $CaCl_2$, and $FeSO_4$ $7H_20$) were obtained from Sigma Chemical, St. Louis, MO. Hach BOD nutrient buffer pillows were purchased from Hach Chemical Company, Loveland, CO. The nutrient buffer was amended with excess phosphate buffer (Ergas and Reuss, 2001). High-purity hydrogen and CO_2 gas were obtained from MG Gas Industries, Malvern, PA. Deionized distilled water from a Barnstead Water Purification Water System (Dubuque, IA) was used for preparation of all solutions.

The following parameters were monitored during the course of the continuous experiments: pH, alkalinity, nitrate, nitrite, total organic carbon (TOC), and dissolved organic carbon (DOC). Nitrate and nitrite were measured using the Standard Method 4500-B (AWWA, 1995) and Hach method 8153, on a spectrophotometer (Model Hach DR 4000U, Loveland, CO.), respectively. The effluent pH was measured with an Orion 720 pH Meter. Alkalinity was determined using a Hach digital titrator according to Standard Method 2130 B (AWWA, 1995).

Dissolved oxygen concentrations were measured with a DO meter (Model YSI 520, YSI, Inc., Yellow Springs, CO). TOC and DOC concentrations were measured using the Hach Method 10324 on a spectrophotometer (Model Hach DR 4000U, Loveland, CO).

4.5 Continuous Operation

The experimental setup for the flow-through studies is shown in Figure 4-4. The reactor was operated in continuous mode by supplying an inorganic nutrient buffer and nitrate solution (Table 4-1).

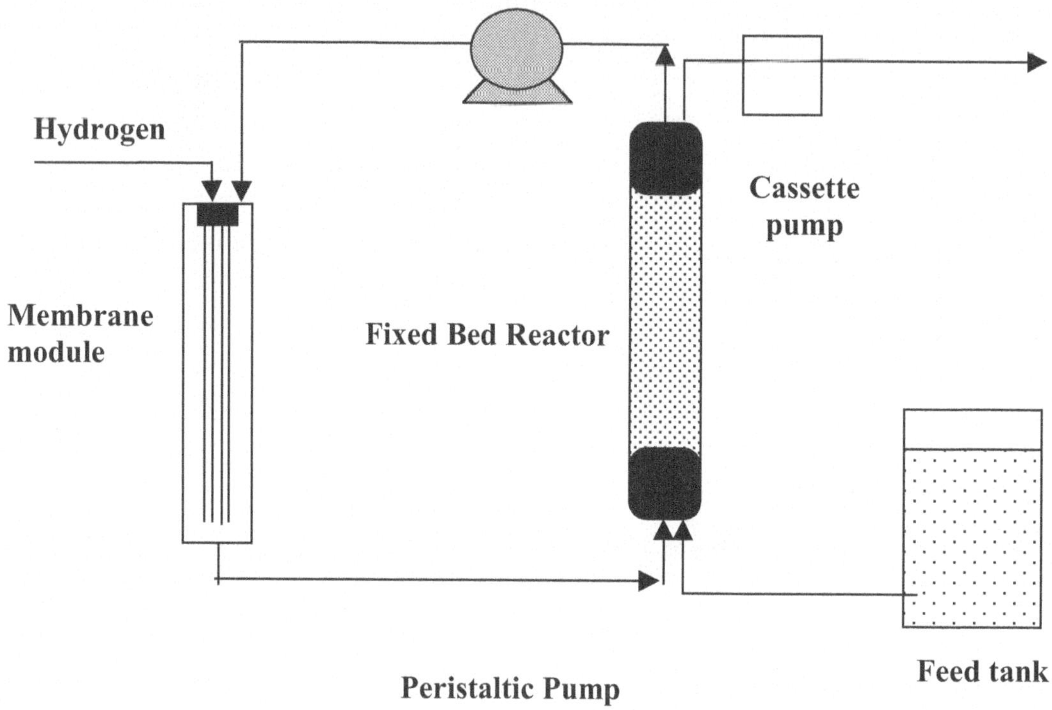

Figure 4-4. Continuous-Flow, Fixed-Film Reactor Setup

Table 4-1. Influent Chemical Composition

Influent Composition	Concentration (mg/L)
$NaNO_3$ as NO_3	100
$NaCO_3$	100
$KHPO_4$	1190
K_2HPO_4	980
$MgSO_4$, $7H_20$	100
$CaCl_2$	30
$FeSO_4$, $7H_2O$	3.75

Influent entered the system from a feed tank that was periodically purged with nitrogen and kept under a nitrogen atmosphere. The system was closed to the atmosphere, with the excess N_2 gas being expelled using check valves at the influent reservoir. The effluent was withdrawn from the top of the fixed-bed bioreactor by a cassette pump (Sarah Standard Cassette Pump, Manostat Corporation, Barrington, IL). The reactor content was continuously recycled through the hollow fiber membrane module by a peristaltic pump (Model 7553-70, Cole-Parmer, Vernon Hills, IL). The operating parameters are listed in Table 4-2. Hydrogen gas was supplied to the sealed end hollow fiber module using a pressure regulator and a pressure sensor.

Table 4-2. Continuous-Flow, Fixed-Film Reactor Parameters

Parameter	Value
Influent flow rate (mL/min)	0.1–0.7
Influent pH	5.85–6.68
Influent alkalinity (mg/L as $CaCO_3$)	380–656
Hydrogen gas pressure (psig)	6.0 and 1.0
Recycle flow rate (L/min)	0.2
Hydraulic retention time (h)	10, 6.75, 3.25, and 1.75

4.6 Results and Discussion

The fixed film reactors were operated at various retention times by gradually increasing the effluent pumping rate. The recycle rate was held constant for a constant hydrogen supply rate. The reported hydraulic retention time (HRT) is obtained by dividing the liquid volume in the system (reactor, membrane module, and tubing) by the influent flow rate. The hydraulic retention time was calculated as follows:

$$HRT = (V_{HFM} + V_r - V_m)/Q_{in}$$

where HRT is the hydraulic retention time in hours, V_{HFM} is the volume of water in the module in liters, V_r is the volume of the reactor column in liters, V_m is the volume of fixed media in liters, and Q_{in} is the influent rate in liters per hour.

The membrane module volume was used in calculating HRT because the recycle ratios were sufficiently high, so the entire system could be considered a completely mixed flow reactor (CMFR). During operation, the membrane surface appeared to be clean; there was no visible biofilm growth on the membrane surface. The steady-state effluent nitrate and nitrite concentrations were used to calculate the removal rates with respect to media volume and media surface area.

The system attained steady state quickly when the operation was switched from batch to continuous flow mode with a 10-h retention time. The influent was buffered adequately; therefore, no pH shift was observed in the effluent, as shown in Figure 4-5. The pH variation range was from 6.5 to 6.65 during steady-state operation. The alkalinity of the reactor effluent also reached steady state after about 200 hours of operation as shown in Figure 4-6 and was consistently higher (by about 130 mg/L) than the influent. This finding was expected because alkalinity is generated during the biological nitrate reduction process. Ergas and Reuss (2001) and Lee and Rittmann (2001) reported similar increases in alkalinity for nitrate reduction experiments.

The observed NO_3^--N concentration in the influent and effluent and NO_2^--N concentration in the reactor effluent are presented in Figure 4-7 as a function of time. At an HRT of 10 hours, complete nitrate removal was observed during steady state. Effluent NO_2^--N concentration in the reactor was 1.54 mg/L. The overall nitrate and nitrogen removal efficiencies were 100% and 93%, respectively, and are shown in Figure 4-9.

The hydrogen pressure in the membrane module was held at 6 psi. Using the membrane module parameters and mass transfer correlation developed earlier in this report, the hydrogen transfer rate of the membrane module at 6 psi operating pressure, was estimated to be 1.31×10^{-3} mg/sec. From reaction stoichiometry, the hydrogen requirement corresponding the reactor nitrate loadings was estimated to be 1.31×10^{-5} mg/sec. Therefore, it is evident that the hydrogen transfer rates were about two orders of magnitude (100 times) greater than the stoichiometric hydrogen requirement, and the process was not expected to be hydrogen limited at this loading condition.

The reactor was also operated at 6.75-h, 3.25-h, and 1.75-h HRTs by increasing the flow rate. The hydrogen pressure in the membrane module was reduced to 1 psi because of hydrogen frequent bubbling problems. The steady state pH, alkalinity, nitrogen species and percentage removals for these HRTs are presented in Figure 4-9 through Figure 4-20. After about 130 hours of operation, at 6.75-h HRT, the influent tank was found to have visible growth of slime layers on reservoir's interior surface. The influent nitrate concentration dropped as a result of this contamination, as shown in Figure 4-11, but no changes in reactor performance or effluent concentrations were observed at this time. The influent tank and the tubing were replaced with cleaned and autoclaved tank and tubing. No other significant operating problems were encountered during the experiments.

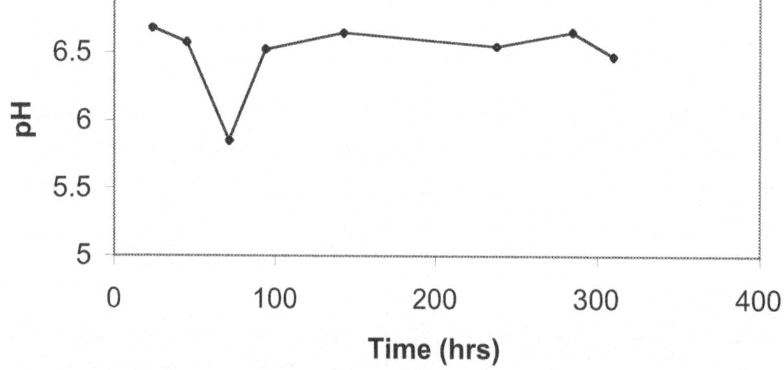

Figure 4-5. Observed Effluent pH during Steady State Operation for 10-h HRT

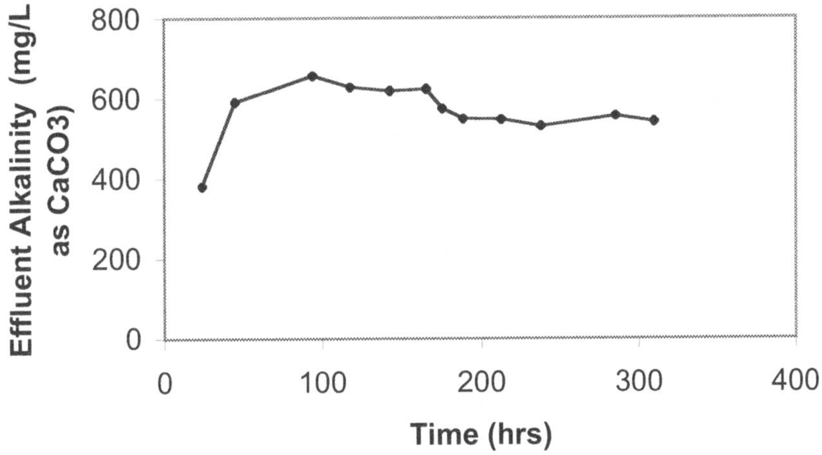

Figure 4-6. Observed Effluent Alkalinity (mg/L as $CaCO_3$) during Steady State Operation for 10-h HRT

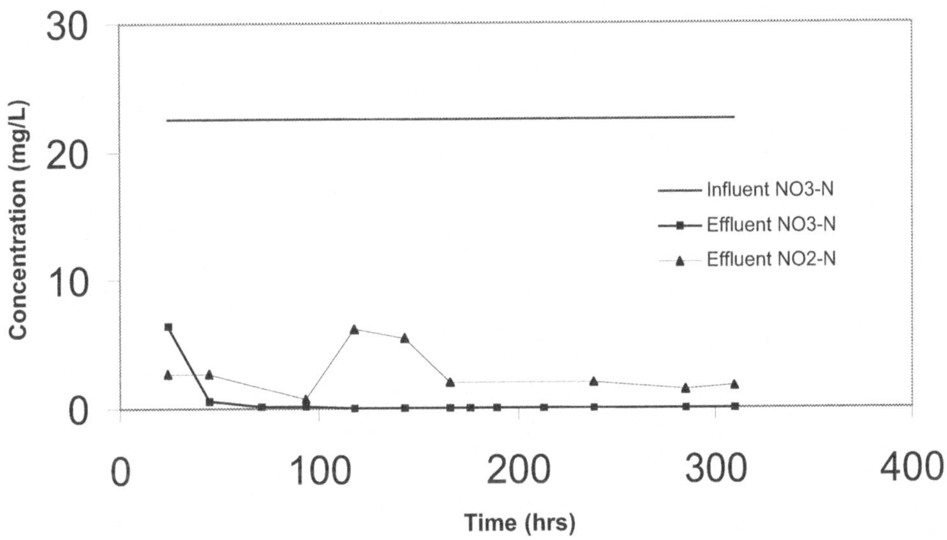

Figure 4-7. Observed NO_3^--N in the Influent and Effluent and NO_2^--N in the Effluent during Steady State Operation for 10-h HRT

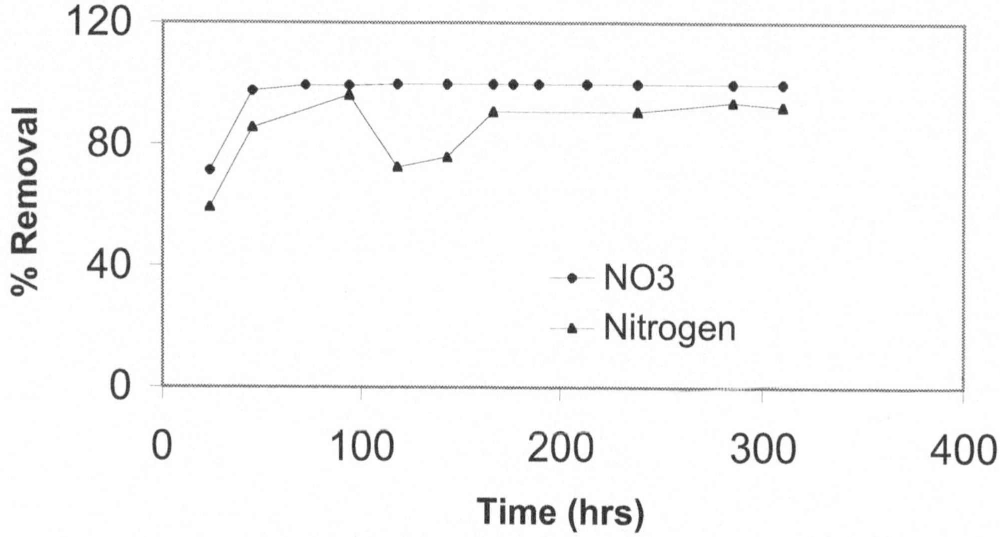

Figure 4-8. Observed NO_3^--N and Total-N Removal Efficiencies during Steady State Operation for 10-h HRT

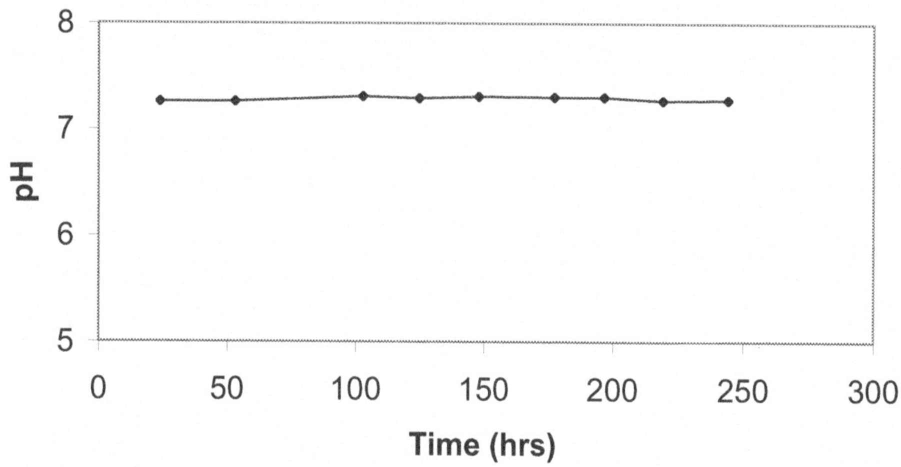

Figure 4-9. Observed Effluent pH during Steady State Operation for 6.75-h HRT

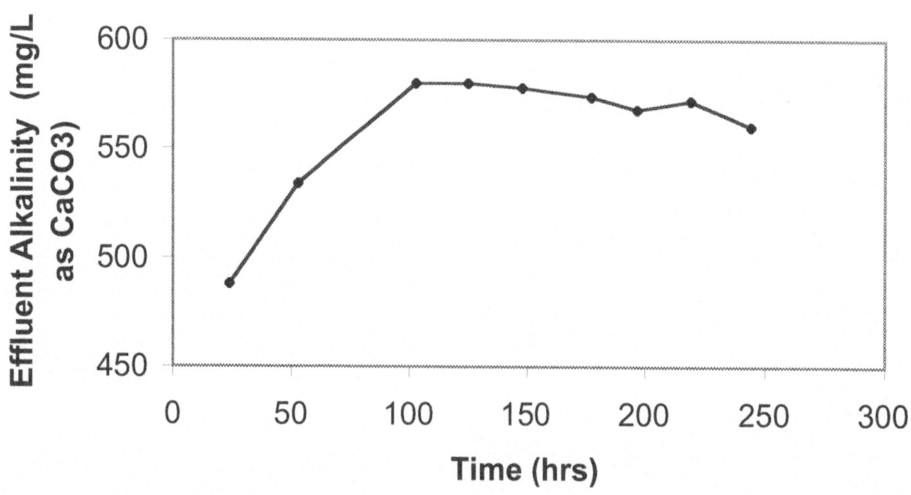

Figure 4-10. Observed Effluent Alkalinity (mg/L as $CaCO_3$) during Steady State Operation for 6.75-h HRT

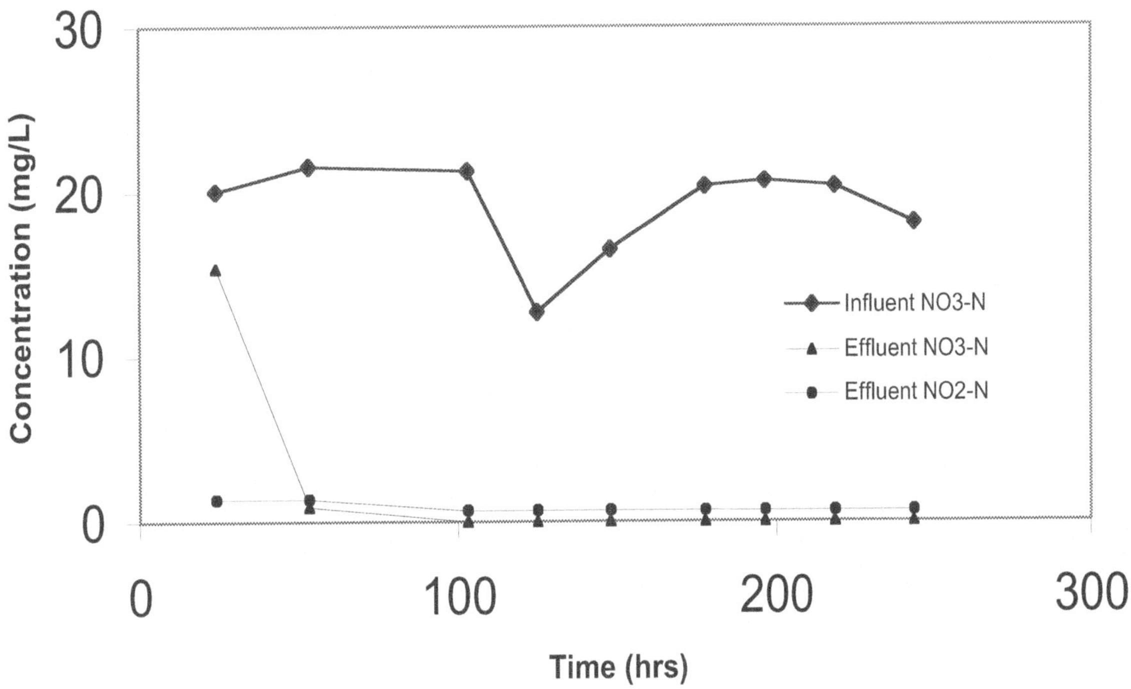

Figure 4-11. Observed NO_3^--N in the Influent and Effluent and NO_2^--N in the Effluent during Steady State Operation for 6.75-h HRT

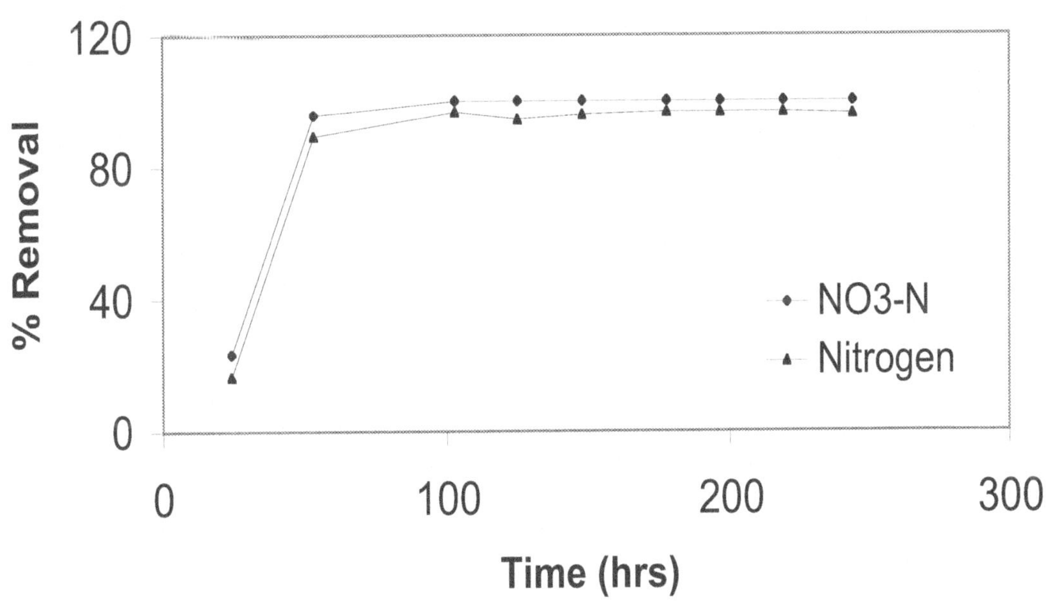

Figure 4-12. Observed NO_3^--N and Total-N Removal Efficiencies during Steady State Operation for 6.75-h HRT

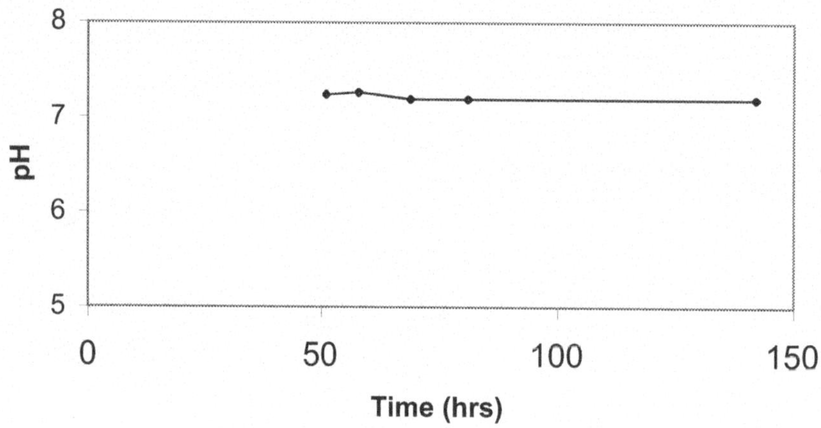

Figure 4-13. Observed Effluent pH during Steady State Operation for 3.25-h HRT

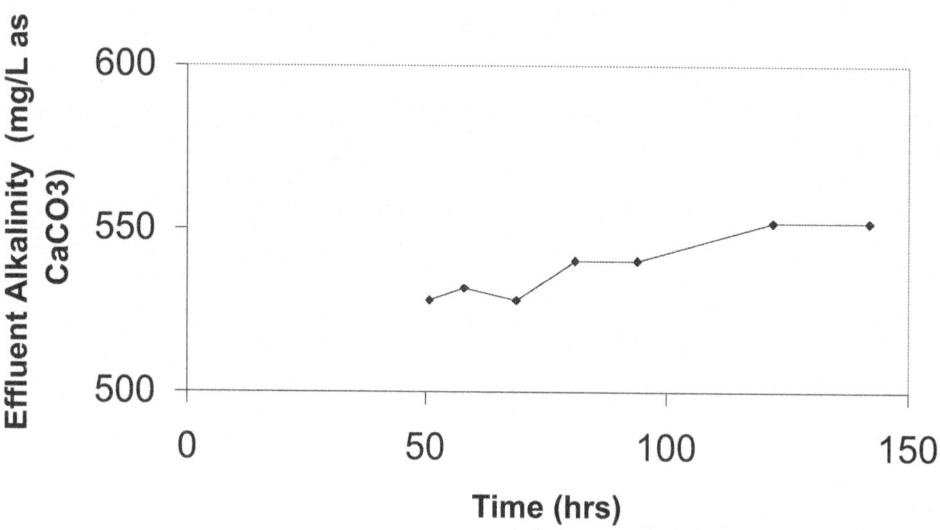

Figure 4-14. Observed Effluent Alkalinity (mg/L as $CaCO_3$) during Steady State Operation for 3.25-h HRT

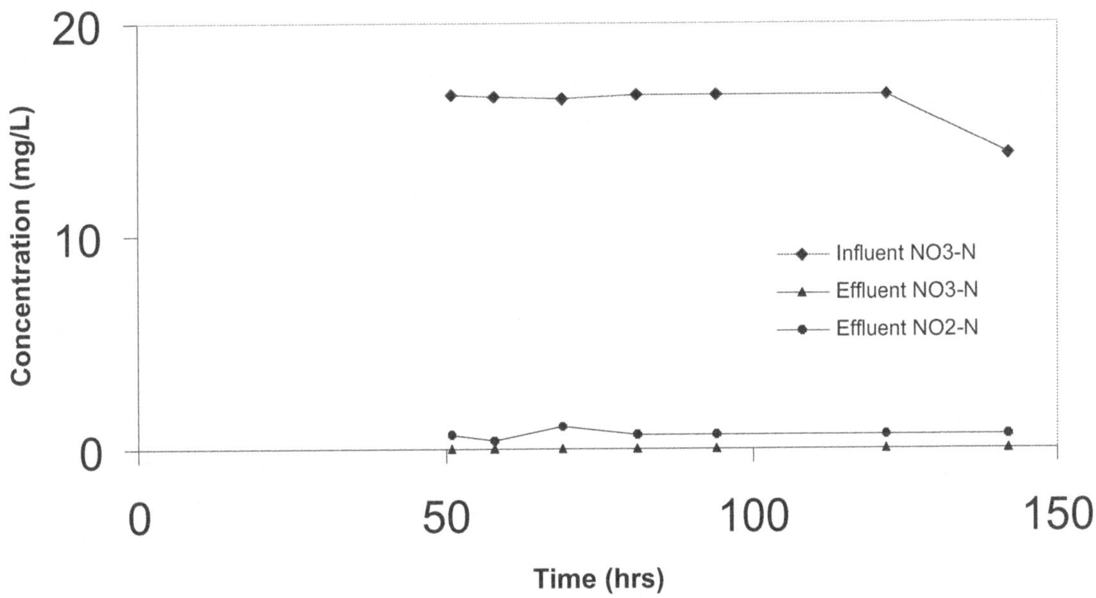

Figure 4-15. Observed NO_3^--N in the Influent and Effluent and NO_2^--N in the Effluent during Steady State Operation for 3.25-h HRT

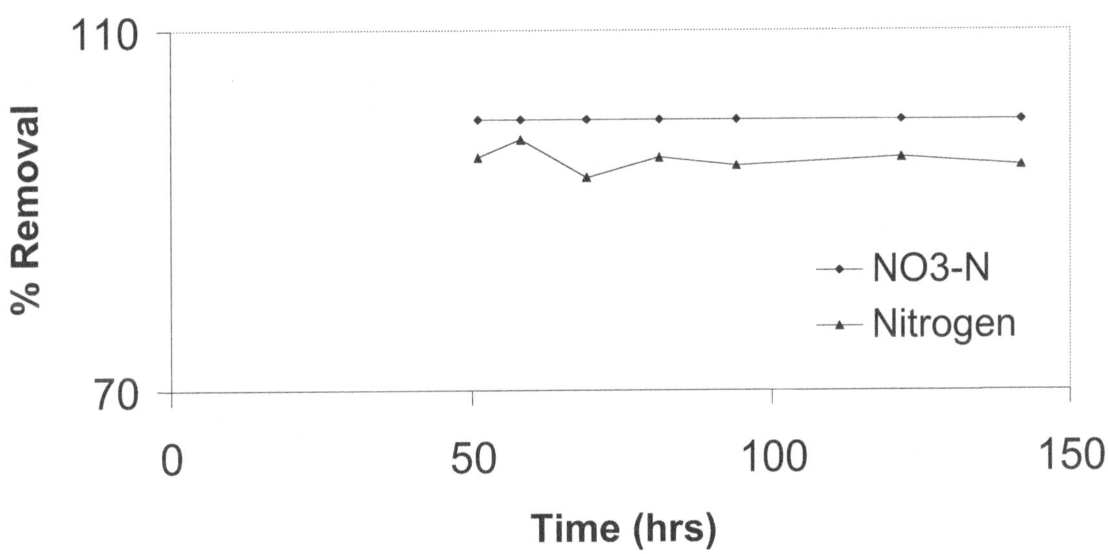

Figure 4-16. Observed NO_3^--N and Total-N Removal Efficiency during Steady State Operation for 3.25-h HRT

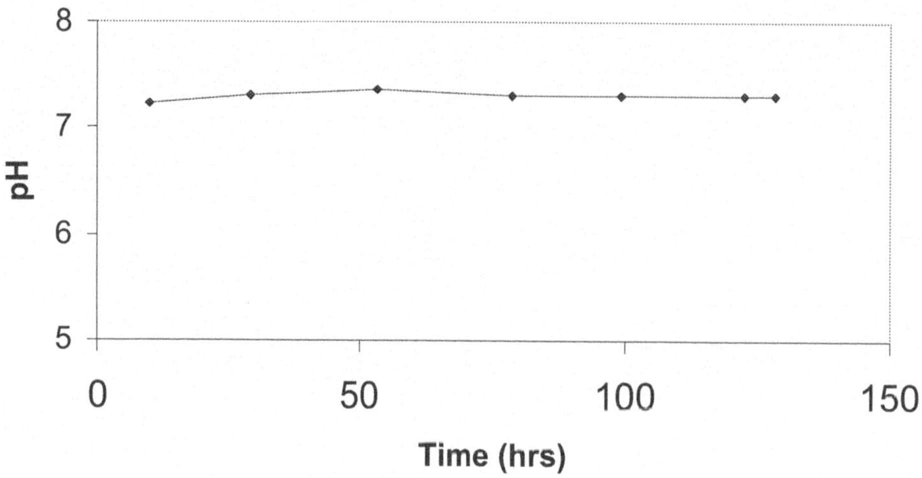

Figure 4-17. Observed Effluent pH during Steady State Operation for 1.75-h HRT

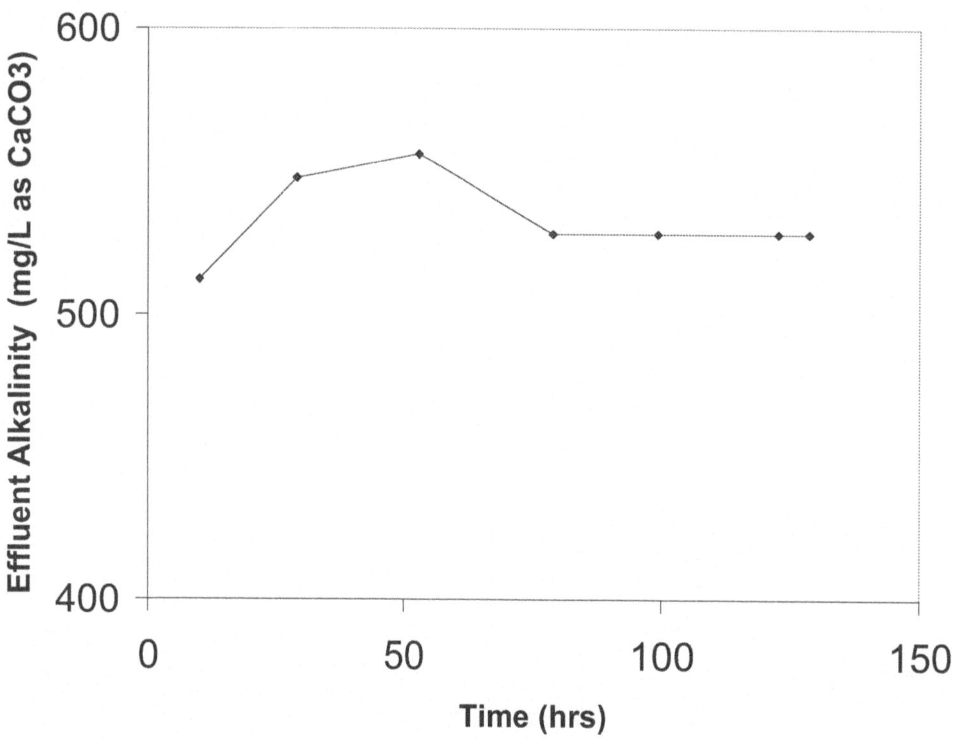

Figure 4-18. Observed Effluent Alkalinity (mg/L as $CaCO_3$) during Steady State Operation for 1.75-h HRT

The reactor operation was discontinued with the appearance of 1.27 mg/L of NO_3^--N at 1.75-h HRT and a concomitant decrease of removal efficiencies as shown in Figure 4-19 and Figure 4-20, respectively. The ratio of hydrogen delivered to stoichiometric hydrogen requirement rate was about 5:1. Therefore, one of the factors that may be limiting at 1.75-h HRT was dissolved hydrogen concentration. It should be noted, however, that increasing the following physical and operating parameters would increase the hydrogen transfer rate from the membrane module:

- length of the fiber and the module;
- number of fibers;
- diameter of the fibers;
- recycle flow rate through the module; and
- operating pressure.

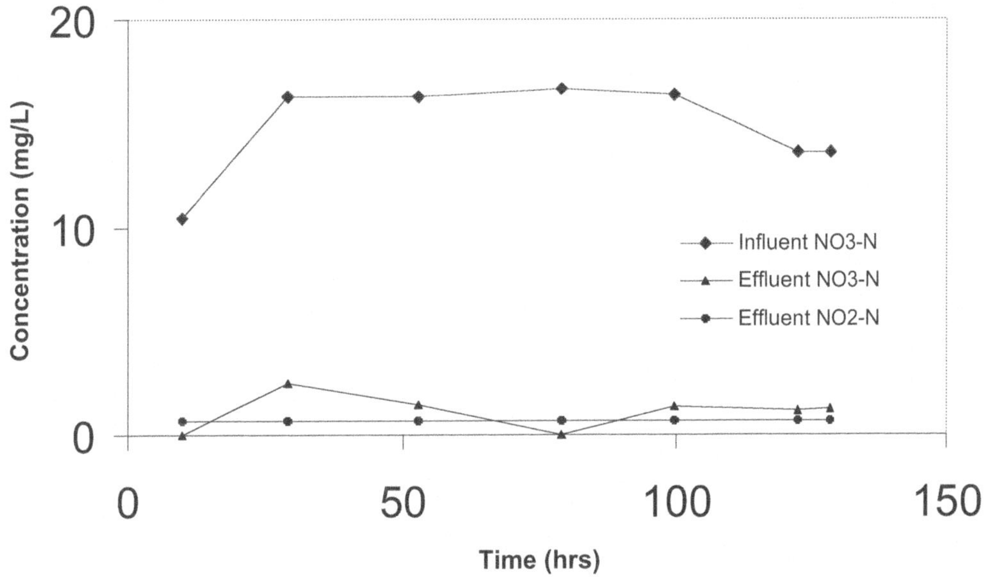

Figure 4-19. Observed NO_3^--N in the Influent and Effluent and NO_2^--N in the Effluent during Steady State Operation for 1.75-h HRT

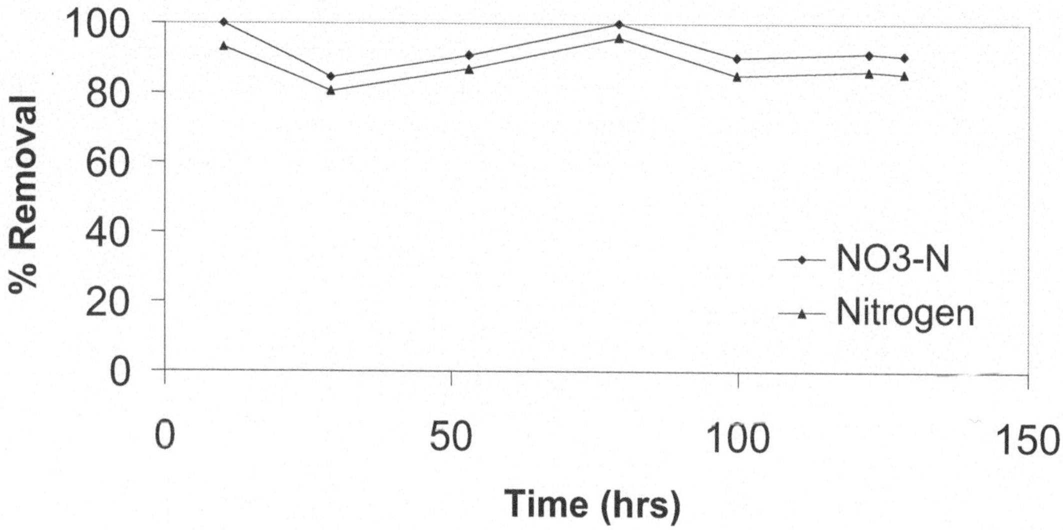

Figure 4-20. Observed NO_3^--N and Total-N Removal Efficiency during Steady State Operation for 1.75-h HRT

The calculated steady state system performance is also summarized in Table 4-3. The table shows no nitrate and low constant nitrite concentrations in the reactor effluent for 10, 6.75 and 3.25 HRTs; this may be an indication that the concentrations of dissolved hydrogen in the reactor were not limiting for the HRTs. Efforts were made to measure the dissolved hydrogen concentration in the reactor effluent by connecting the effluent line to the Orbisphere hydrogen sensor. The dissolved hydrogen concentration varied from 30 to 40 ppb at 1 psi operating pressure. Effluent dissolved hydrogen concentration was not measured at 6 psi operating pressure.

Table 4-3. Overall Steady State Performance of Fixed-Film, Continuous-Flow Membrane Bioreactor

HRT (h)	Influent	Effluent					% Removal	
	NO_3^--N (mg/L)	pH	Alkalinity (mg/L as $CaCO_3$)	NO_3^--N (mg/L)	NO_2^--N (mg/L)	Total-N (mg/L)	NO_3^--N	Total-N
10	22.6*	6.5	556	0	1.54	1.54	100	93.1
6.75	20.46	7.2	565	0	0.68	0.68	100	96.67
3.25	16.64	7.2	552	0	0.68	0.68	100	95.8
1.75	13.58	7.3	528	1.27	0.68	1.95	90.6	85.5

*Calculated

The steady state volumetric NO_3^--N loading rate and removal efficiencies per unit surface area for each HRT was calculated and presented in Table 4-4. The data show that the optimum performance with little (about 0.68 mg/L) nitrite concentration would be achieved between 1.75 and 3.25-h HRT.

Table 4-4. Continuous-Flow, Fixed-Film Reactor Steady State Results

HRT (hr)	Volumetric Loading (mg NO_3^--N /m^3-h)	Removal Efficiency per Unit Surface Srea (mg NO_3^--N /m^2-h)	Effluent Nitrate Concentration (mg NO_3^--N/L)	Effluent Nitrite Concentration (mg NO_2^--N/L)
1.75	12,978	13.21	1.27	1.95
3.25	6906	7.03	0	0.68
6.75	3335	3.40	0	0.68
10	2263	2.30	0	1.54

The reported loading rates and the corresponding effluent concentrations may be used for scale up purposes for similar systems. The reactor performance was compared with other similar studies in the literature and is presented in Table 4-5.

Table 4-5. Continuous-Flow, Fixed-Film Reactor Performance

	% Removal	HRT (hr)	Surface area removal efficiency (mg NO_3^--N/ m^2-h)	Volumetric removal rate (kg NO_3^--N/ m^3-day)
Benedict et al. (1996)	-	6	42.7	2.63
Lee and Rittmann (2000)	76, 92	0.7	33, 42	2.6, 3.95
Ergas and Reuss (2001)	90	30	104	0.059
This study	93	1.75	13.2	0.312

The recycle ratios were determined by dividing the flow into the reactor by the recycle flow rate through the membrane reactor (Table 4-6). Recycle ratio may be an operational concern because it is much higher than those recommended for trickling filters (Metcalf and Eddy, 1991). Recycle rates provide control of oxygen transfer rates and biofilm thickness on the media in typical trickling filter operations. Higher recycle rates would necessitate higher pumping requirements, thereby increasing the operating cost. However, the recycle ratios were found to be within the range of similar studies employing autotrophic denitrification reactors (Lee and Rittmann, 2000; Ergas and Reuss, 2001; Benedict et al., 1996), as shown in Table 4-6.

Table 4-6. Comparison of Recycle Ratios from Literature

	System Configuration	Recycle Ratio
Metcalf and Eddy (1991)	Trickling filters	0.5–2
Lee and Rittmann (2000)	Hollow fiber membrane bioreactor	175
Ergas and Reuss (2001)	Hollow fiber membrane bioreactor	125–755
Benedict et al. (1996)	Fixed film bioreactor	114–343
This study	Fixed film bioreactor	143–500

4.7 Conclusions

A stable biofilm was developed in the packed bed reactor that removed nitrate using hydrogen as the electron donor. Hydrogen gas was successfully delivered to the reactor via the hollow fiber membrane gas transfer module without any visible biological fouling. The calculated hydrogen transfer rate of the given module indicated that the system did not experience hydrogen limitations up to a HRT of about 3.25 h.

CHAPTER 5.0

CONCLUSIONS AND RECOMMENDATIONS

5.1 Conclusions

The main goal of this project was to evaluate the feasibility of a continuous-flow, fixed-film bioreactor coupled with a hollow fiber membrane module for autotrophic denitrification. Design relationships for hydrogen gas transfer using hollow fiber membranes were developed and the effect of HRT and hydrogen transfer on nitrate removal in the system was evaluated. Batch studies were conducted to develop acclimated mixed cultures capable of autotrophic denitrification. Sodium carbonate and CO_2 were delivered to the batch reactors for delivery of inorganic carbon.

The following conclusions can be drawn from this project:

1. Autotrophic denitrifying hydrogen oxidizing bacteria can be easily isolated from wastewater treatment plants with biological nutrient removal facilities;
2. No differences were observed when inorganic carbon was delivered via Na_2CO_3 and CO_2;
3. Some nitrite accumulations were observed in the batch studies; however, nitrite was removed during the duration of the experiments;
4. Gas transfer studies indicated that hydrogen transfer is primarily controlled by liquid film diffusion;
5. The following mass transfer correlation can be used to design membrane modules for hydrogen gas transfer to water:

   $$Sh = 2.68\ Re\,de/L^{1.02}\ Sc^{0.33}$$

   where:

   Sh is the Sherwood number;
   Re is the Reynolds number;
   Sc is the Schmidt Number;
   de is the equivalent diameter of the membrane; and
   L is the length of the membrane.

6. Continuous flow studies indicated that a stable biofilm can be developed in a packed bed reactor to remove nitrate, using hydrogen as the electron donor;
7. Hydrogen gas could be successfully and safely delivered to the reactor via the hollow fiber membrane gas transfer module without visible biological fouling;
8. Hydrogen concentrations indicated that the system did not experience hydrogen limitations;
9. No significant nitrite accumulation was observed in system operation; and
10. Nitrate removal and hydrogen utilization efficiencies compared favorably to other studies.

5.2 Suggested Future Work

5.2.1 Evaluation of Long-Term System Performance

Several attached growth biofilm reactors can be operated for extended periods with membrane modules supplying the required gases for evaluating long-term performance of the membrane modules. Fouling was not a problem during the flow-through studies reported in Chapter 4.0. Nevertheless, reactor performance should be assessed over a longer time. Efforts need to be made to identify and address operating problems. Of major concern is biological fouling caused by biofilm growth on the membranes. The required maintenance procedures can then be developed for the gas transfer modules. The cleaning frequency and module lifespan need to be determined.

5.2.2 Pilot Plant Study

This study focused on laboratory-scale experiments for hydrogen gas delivery for autotrophic denitrification. Pilot-scale studies need to be conducted for hollow fiber membrane denitrification systems.

GLOSSARY

Autotrophic Organisms that derive cell carbon from inorganic sources, such as carbon dioxide.

Heterotrophic Organisms that derive cell carbon from organic sources.

Batch Reactor A reactor that has no flow (that is, flow neither enters nor leaves the reactor).

Denitrification The biological reduction of nitrate to nitrogen gas by denitrifying bacteria.

Hollow fiber membranes These are small, thin-walled membranes that consist of capillary or large-bore hollow fibers of diameters ranging from 42 μm for reverse osmosis to 1100 μm for ultrafiltration. The major advantage of hollow fiber membrane systems is that large numbers of fibers can be packed inside a compact cartridge, yielding area-to-volume ratios from 600 to 1200 m^2/m^3.

Hydraulic retention time (HRT) The time obtained by diving the empty volume of a reactor by its flow rate.

Membranes Membranes are primarily used for separation, and membrane processes are generally separating processes. A membrane process requires two bulk phases physically separated by a third phase, the membrane. The membrane is an interphase between the two bulk phases. The membrane phase interposed between two bulk phases controls the exchange of mass between the two bulk phases in a membrane process.

Nonpoint source pollution Pollutants emanating from an unconfined or unchannelled source, including agricultural runoff, drainage or seepage and air contamination from landfills or surface impoundments.

REFERENCES

Ahmed. T., and M. J. Semmens.1992a. Use of sealed end hollow fibers for bubbleless membrane aeration: Experimental studies. *J. Membrane Science* 69:1-10.

Ahmed. T. ,and M. J. Semmens.1992b. The use of independently sealed microporous hollow fiber membranes for bubbleless oxygenation of water: model development. *J. Membrane Science* 69:11-20.

American Water Works Association (AWWA). 1995. Surveys reveal pesticides, nitrates in water from farming areas. *J. American Water Works Association* 87(10):13-14.

Barrenstein, A., U. Kramer, and P. Obermann. 1986. Underground treatment of nitrate-rich groundwater by infiltration with treated wastewater or methane-rich natural gas. *DVGW-Schriftenreihe*. Wasser. Frankfurt, West Germany. 106:99-116.

Benedict, S.W. 1996. Denitrification with autotrophic hydrogen oxidizing bacteria. *M.S. thesis*. University of Nevada. Reno, NV.

Benedict, S.W., T. Ahmed, and K. Jahan .1998. Autotrophic denitrification using hydrogen oxidizing bacteria in continuous flow biofilm reactor. *Toxicological and Environmental Chemistry* 67:197-214.

Bennett, C.O., and J.E. Myers.1982. *Momentum, Heat and Mass Transfer*. 3rd Ed. New York: McGraw-Hill.

Brautigam,. H.J., and J. Sckoulov. 1987. Denitrification with hydrogen as an electron donor introduced through nonporous membranes. *Ber. Abwassertech. Ver* 37:555-65.

Brezonik, P.L.1977. Denitrification in natural waters. *Prog. Wat. Tech*. 8(4-5):373-392.

Brindle, K., T. Stephenson, and M.J Semmens.1998. Nitrification and oxygen utilization in a membrane aeration bioreactor. *J. Membrane Science* 144:197.

Chang, C.C., S.K. Tseng, and H.K. Huang. 1999. Hydrogenotrophic denitrification with immobilized *Alcogenes eutrophus* for drinking water treatment. *Bioresource Technology* 69:53-58.

Claus, G., and H.J. Kutzner. 1985. Autotrophic denitrification *Thiobacillus denitrificans*. *Appl. Microbiology and Biotechnology* 22: 289-296.

Clifford, D., and X. Liu. 1993. Ion exchange for nitrate removal. *J. American Water Works Association* 85(4):135-143.

Cote, P.L., J. Bersillon, and A. Huyard.1989. Bubble-free aeration using membranes: Mass transfer analysis. *J. Membrane Science* 47:91-106.

Croll, B.T. 1985. Biological fluidized-bed denitrification for potable water. In *Proc. International Symposium on Advances in Water Engineering.* T.H.Y. Tebbut, ed. London: Elsevier Applied Sciences. 180-187.

Croll, B.T., and C.R. Hayes. 1988. Nitrate and water supplies in the United Kingdom. *Envir. Pollution* 50:163-187.

Delanghe, B., F. Nakamura, H. Myoga, and Y. Magara. 1994a. Biological Denitrification with Ethanol in a Membrane Bioreactor. *Environmental Technology* 15: 61-70.

Delanghe, B., J. Nakamura,H. Myoga, Y. Magara, and E. Guibal .1994b. Drinking water denitrification in a membrane bioreactor. *Water Science Technology* 30(6):157-160.

Dries, D., J. Liessens, W. Verstrate, P. Stevens, P. de Vos, and J. Ley.1988. Nitrate removal from drinking water by means of hydrogenotrophic denitrifiers in a polyurethane carrier reactor. *Water Supply* 6:181-192.

Eckenfelder, W. W., Jr. .1989. *Industrial Water Pollution Control*. 2nd Ed. New York: McGraw-Hill.

Environmental Protection Agency (EPA) .1993. *Manual Nitrogen Control.* EPA/625/R-93/010. September.

Environmental Protection Agency (EPA). 1998. *Technical Protocol for Evaluating Natural Attenuation of Chlorinated Solvents in Groundwater.* EPA/600/R-98/128. September.

Ergas, S.J., and A.F. Reuss. 2001. Hydrogenotrophic denitrification of drinking water using a hollow fibre membrane bioreactor. *J. Water Supply: Res. Technol.-AQUA* 50(3):161-171.

Freeze, R.A., and J.A. Cherry. 1979. *Groundwater.* Englewood Cliffs, N J: Prentice Hall.

Fuchs, W., G. Schatzymayr, and R. Braun. 1997. Nitrate removal from drinking water using a membrane-fixed biofilm reactor. *Appl. Microbial Biotechnology* 48:267-274.

Gantzer, C.J. 1995. Membrane dissolution of hydrogen for biological nitrate removal. In *Proceedings of the 68th Annual Conference and Exhibition. Water Environment Federation Research Symposium.* Water Environment Federation. Miami Beach, FL. pp. 49-60.

Gayle, B.P., G.D. Boardman, J.H. Sherrard, and R.E. Benoit .1989. Biological denitrification of water. *J. Environmental Engineering* 115(5):930-943.

Gross, H., and K. Treutler.1986.. Biological denitrification process with hydrogen oxidizing bacteria for drinking water treatment. *Aqua* 5:288-290.

Gross, H., G. Schnoor, and P. Ruther. 1988. Biological denitrification process with hydrogen oxidizing bacteria for drinking water treatment. *Water Supply* 6:193.

Hallberg, G. R. 1989. Nitrate in ground water in the United States. In *Nitrogen Management and Ground Water Protection. Developments in Agriculture and Managed-Forest Ecology.* New York: Elsevier.

Hoerner, G., and A. Irmer .1989. Autotrophic biological denitrification with the membrane fluidized-bed technology with hydrogen as electron donor. *Vom Wasser* 73:429-47.

Islam, S., and M. T. Suidan.1994. Enhancement and control of denitrification using a biofilm electrode reactor. In *Proceedings of the Water Environment Federation Annual Conference*. Chicago. pp 217-225.

Iversen, S. 1995. Absorption or stripping using hollow-fiber membranes. .FLS Miljoe A/S. Den. PCT Int. Appl. 83.

Jimenez, K.M., A.K. Zander, and S.J. Grimberg .1998. Membrane enhanced biofilm reactor for treatment of high strength wastewaters containing nitrophenilic compounds." In *Proceedings of the Water Environmental Federation. 71st annual Conference Exposition.* Orlando, FL.

Kapoor. A., and T. Viraraghavan. 1997. Nitrate removal from drinking water: Review. *ASCE J. of Environmental Engineering* 123(4):371-380.

Knudsen, J.G., and D.L. Katz. 1958. *Fluid Dynamics and Heat Transfer.* New York: McGraw-Hill.

Korom, S. 1992. Natural denitrification in the saturated zone: A review. *Water Research* 28 (6):1657-1668.

Kurt.,M., J. Dunn, and J.R. Bourne. 1987. Biological denitrification of drinking water using autotrophic organisms with hydrogen in a fluidized-bed reactor. *Biotechnology and Bioengineering* 29:493-501.

Lazarova, V.Z., B. Capdeville, and L. Nikolov. 1992. Biofilm performance of a fluidized bed biofilm reactor for drinking water denitrification. *Water Science Technology* 26 (3-4):555-566.

Lee, K-C., and B.E. Rittmann. 2000. A novel hollow-fibre membrane biofilm reactor for autohydrogenotrophic denitrification of drinking water. *Water Science Technology* 14 (4-5): 219-226.

Li, K. 2000. Design of hollow fibre membrane modules for soluble gas removal. *Chemical Engineering Science* 55(23): 5579-88.

Liessens, J., R. Germonpre, I. Kersters, S. Beernaert, and W. Verstrate.1993. Removing nitrate with a methylotrophic fluidized bed: Microbiological quality. *J. American Water Works Association* 85(4): 155-161.

Lowry,O. H., N. J.Rosebrough, A.L.Farr, and R.J. Randall. 1951. Protein measurements with the Folin phenol reagent. *J. Biol. Chem.* 193: 265-275

Mateju, V., S. Cizinska, J. Kreici, and T. Janoch. 1992. Biological water denitrification: A review. *Enzyme Microbial Technology* 14:170-183.

Metcalf & Eddy, Inc. 1991. *Wastewater Engineering: Treatment, Disposal, and Reuse*. 3rd Ed. New York: McGraw Hill.

Miyake, Y., H. Myoga, and Y. Magara. 1997. Several factors affecting the specific denitrification rate of hydrogen oxidizing bacteria. *Water Use Conservation Mex. '96 Workshop.*

Mueller, J.A., and W.C. Boyle, 1988. Oxygen transfer under process conditions. *J. WPCF* 60: 332-341.

Novak, P.J., J.A. Edstrom, L.W.. Clapp, R.M. Hozalski, and M.J. Semmens. 2001. Stimulation of dechlorination by membrane delivered hydrogen: Small field demonstration. In *Proc. 6th International Symposium on In-Situ and On-site Bioremediation.* San Diego.

Oh, S.-E., K.-S. Kim, H.-C. Choi, J. Cho, and I. S. Kim. 2000. Kinetics and physiological characteristics of autotrophic denitrification by denitrifying sulfur bacteria. *Water Sci. Technol.* 42: 3-4.

Pekdemir, T., E.K. Kacmazoglu, B. Keskilner, and O.F. Algur. 1998. Drinking water denitrification in a fixed bed packed biofilm reactor. *Turkish Journal of Engineering and Environmental Sciences* 22:39-45.

Rittmann, B.E., and P.M. Huck. 1989. Biological treatment of public water supplies. *CRC Critical Reviews on Environmental Engineering Control* 119(2):119-184.

Sakakibara, Y., and J. Kusaka. 1999. "In situ autotrophic denitrification using electrode under oligotrophic conditions." In *Proceedings of the 5th International In Situ On-Site Bioremediation Symposium.* 4:73-78. Columbus, OH: Battelle Press.

Schmidt, I., J. Beyersdore, U. Driks, P. Remmers, and K.D. Vorlop. 1989. Development of a process for denitrification of drinking water. *Dechema Biotechnology Conference.* 3, 833-836.

Selenka, F., and R. Dressler. 1990. Microbiological and chemical investigations on a biological, autotrophic denitrification plant using hydrogen as an energy source. *J. Water SRT-Aqua* 39(2): 107-116.

Sell, M., H.. Ross, R. FeistBehling, K. Kneifel, and K.V. Peinemann. 1993. Hydrogen transport into water via hollow fiber membranes for catalytic nitrate removal. In *Proceedings of the Membrane Technology Conference*. Denver, CO.: American Water Works Assoc. pp. 471-90.

Semmens, M. J. 1991. *Bubbleless gas transfer device and process.* U.S. Patent # 5,034,164.

Shuvai, H.I., and V. Gruener. 1977. Infant methomoglobinemia and other health effects of nitrates in drinking water. *Progress in Water Technology* 8(4-5):183.

Smith, R.L., M.L Ceazan, and M.H. Brooks. 1994. Autotrophic hydrogen oxidizing denitrifying bacteria in groundwater: Potential agents for bioremediation of nitrate contamination. *Applied and Environmental Microbiology* 60(6):1949-1955.

Szekeres, S., I. Kiss, T. Bejerano, and M.I. Soarez. 2001. Hydrogen-dependent denitrification in a two-reactor bio-electrochemical system. *Water Res.* 35(3):715-719.

Trouve, C., and P.M. Chazal. 1999. Autotrophic denitrification by *Thiobacillus denitrificans* with thiosulfate as sole energy source. Pilot scale experiments at low temperature. *Environ. Technol.* 20(4):387-395.

Urbain,V., R. Benoid, and J. Manem .1996. Membrane bioreactor: a new treatment tool. *J. American Water Works Association* 88:75-86.

van der Hoek, J.P., W.F. van der Hoek, and A. Klapwijk. 1988. Nitrate removal from groundwater: Use of a nitrate selective resin and a low concentrated regenerant. *Water Air and Soil Pollution* 37:41.

van Houten, R., L.W. Hulshoff, and G. Lettinga. 1994. Biological sulfate reduction using gas-lift reactors fed with hydrogen and carbon dioxide as energy and carbon source. *Biotechnology and Bioengineering* 44:586-594.

Viraraghavan, T., and G.A.K. Rao. 1990. Treatment processes for nitrate removal from water. In *Proceedings of the 42nd Annual Convention of Western Canada Water and Wastewater Association*. Regina, Saskatchewan, Canada. 137-160.

Weber, W.J., and F.A. DiGiano. 1996. *Process Dynamics in Environmental Systems.* New York: John Wiley & Sons.

Yang, M. C., and E.L. Cussler. 1986. Designing hollow-fiber contactors. *AIChE Journal* 32(11): 1910-1916.

WASTEWATER UTILITY

Alabama
Montgomery Water Works & Sanitary Sewer Board

Alaska
Anchorage Water & Wastewater Utility

Arizona
Gila Resources
Glendale, City of, Utilities Department
Mesa, City of
Phoenix Water Services Department
Pima County Wastewater Management

Arkansas
Little Rock Wastewater Utility

California
Calaveras County Water District
Central Contra Costa Sanitary District
Contra Costa Water District
Crestline Sanitation District
Delta Diablo Sanitation District
Dublin San Ramon Services District
East Bay Municipal Utility District
El Dorado Irrigation District
Escondido, City of
Fairfield-Suisun Sewer District
Irvine Ranch Water District
Las Virgenes Municipal Water District
Lodi, City of
Los Angeles, City of
Los Angeles County, Sanitation Districts of
Napa Sanitation District
Orange County Sanitation District
Palo Alto, City of
Riverside, City of
Sacramento Regional County Sanitation District
San Diego Metropolitan Wastewater Department, City of
San Francisco, City & County of
San Jose, City of
Santa Barbara, City of
Santa Rosa, City of
South Bayside System Authority
South Orange County Wastewater Authority
Union Sanitary District
West Valley Sanitation District

Colorado
Boulder, City of
Colorado Springs, City of
Littleton/Englewood Water Pollution Control Plant
Metro Wastewater Reclamation District, Denver

Connecticut
New Haven, City of, WPCA
District of Columbia
District of Columbia Water & Sewer Authority

Florida
Broward, County of
Fort Lauderdale, City of
Gainesville Regional Utilities
JEA
Kissimmee, City of, Department of Water Resources
Miami-Dade Water & Sewer Authority
Orange County Utilities Department
Orlando, City of
Reedy Creek Improvement District
Seminole County Environmental Services
St. Petersburg, City of
Stuart Public Utilities
Tallahassee, City of
Tampa, City of
West Palm Beach, City of

Georgia
Atlanta Department of Watershed Management
Augusta, City of
Clayton County Water Authority
Cobb County Water System
Columbus Water Works
Fulton County
Gwinnett County Department of Public Utilities
Macon Water Authority
Savannah, City of

Hawaii
Honolulu, City and County of

Idaho
Pocatello, City of

Illinois
American Bottoms Wastewater Treatment Plant
Dupage, County of
Greater Peoria Sanitary District
Kankakee River Metropolitan Agency
Metropolitan Water Reclamation District of Greater Chicago
North Shore Sanitary District
Wheaton Sanitary District

Indiana
Fort Wayne, City of

Iowa
Ames, City of
Des Moines Metro Wastewater Reclamation Authority

Kansas
Johnson County Unified Wastewater Districts
Unified Government of Wyandotte County/Kansas City, City of

Kentucky
Louisville & Jefferson County Metropolitan Sewer District

Louisiana
Sewerage & Water Board of New Orleans

Maine
Bangor, City of
Portland Water District

Maryland
Anne Arundel County Bureau of Utility Operations
Howard County Department of Public Works
Washington Suburban Sanitary Commission

Massachusetts
Boston Water & Sewer Commission
Upper Blackstone Water Pollution Abatement District

Michigan
Ann Arbor, City of
Detroit, City of
Grand Rapids, City of
Holland Board of Public Works
Lansing, City of
Owosso Mid-Shiawassee County WWTP
Saginaw, City of
Wayne County Department of Environment
Wyoming, City of

Minnesota
Rochester, City of

Missouri
Independence, City of
Kansas City Missouri Water Services Department
Little Blue Valley Sewer District
Metropolitan St. Louis Sewer District

Nebraska
Lincoln Wastewater System

Nevada
Henderson, City of
New Jersey
Passaic Valley Sewerage Commissioners

New York
New York City Department of Environmental Protection

North Carolina
Cary, Town of
Charlotte/Mecklenburg Utilities
Durham, City of
Metropolitan Sewerage District of Buncombe County
Orange Water & Sewer Authority

Ohio
Akron, City of
Butler County Department of Environmental Services
Columbus, City of
Metropolitan Sewer District of Greater Cincinnati
Northeast Ohio Regional Sewer District
Summit, County of

Oklahoma
Tulsa, City of

Oregon
Clean Water Services
Eugene/Springfield Water Pollution Control
Water Environment Services

Pennsylvania
Philadelphia, City of
University Area Joint Authority, State College

South Carolina
Charleston Commissioners of Public Works
Mount Pleasant Waterworks & Sewer Commission
Spartanburg Sanitary Sewer District

Tennessee
Cleveland, City of

Knoxville Utilities Board

Murfreesboro Water & Sewer Department

Nashville Metro Water Services

Texas
Amarillo, City of

Austin, City of

Dallas Water Utilities

Denton, City of

El Paso Water Utilities

Fort Worth, City of

Gulf Coast Waste Disposal Authority

Houston, City of

San Antonio Water System

Trinity River Authority

Utah
Salt Lake City Corporation

Virginia
Alexandria Sanitation Authority

Arlington, County of

Fairfax County Virginia

Hampton Roads Sanitation District

Henrico, County of

Hopewell Regional Wastewater Treatment Facility

Loudoun County Sanitation Authority

Lynchburg Regional WWTP

Prince William County Service Authority

Richmond, City of

Washington
Edmonds, City of

Everett, City of

King County Department of Natural Resources

Seattle Public Utilities

Yakima, City of

Wisconsin
Green Bay Metro Sewerage District

Madison Metropolitan Sewerage District

Milwaukee Metropolitan Sewerage District

Racine, City of

Sheboygan Regional Wastewater Treatment

Wausau Water Works

Australia
South Australian Water Corp.

Sydney Water Corp.

Water Corp. of Western Australia

Canada
Greater Vancouver Regional District

Toronto, City of, Ontario

Mexico
Servicios de Agua y Drenaje de Monterrey, I.P.D.

United Kingdom
Yorkshire Water Services Limited

STORMWATER UTLITY

California
Los Angeles, City of, Department of Public Works

Monterey, City of

Santa Rosa, City of

Colorado
Boulder, City of

Georgia
Griffin, City of

Kansas
Overland Park, City of

Kentucky
Louisville & Jefferson County Metropolitan Sewer District

Maine
Portland Water District

North Carolina
Cary, Town of

Charlotte, City of, Stormwater Services

Pennsylvania
Philadelphia, City of

STATE

Fresno Metropolitan Flood Control District, Calif.

Urban Drainage & Flood Control District, Colo.

CORPORATE

ADS Environmental Services

The ADVENT Group Inc.

Alan Plummer & Associates

Alden Research Laboratory

Alpine Technology Inc.

Aquateam–Norwegian Water Technology Centre A/S

BaySaver Inc.

BioVir Inc.

Black & Veatch

Boyle Engineering Corporation

BPR CSO

Brown & Caldwell

Burns & McDonnell

CABE Associates Inc.

The Cadmus Group

Camp Dresser & McKee Inc.

Carollo Engineers Inc.

Carpenter Environmental Associates Inc.

CDS Technologies Inc.

Chemtrac Systems Inc.

CH2M HILL

Clancy Environmental Consultants Inc.

Damon S. Williams Associates, LLC

David L. Sheridan, P.C.

Earth Tech Inc.

Ecolab Water Care Services

EMA Inc.

The Eshelman Company Inc.

Finkbeiner, Pettis, & Strout Inc.

Frontier Geosciences Inc.

ftn Associates Inc.

Gannett Fleming Inc.

GE Betz

Golder Associates Inc.

Greeley and Hansen LLC

The HACH Company

Hazen & Sawyer, P.C.

HDR Engineering Inc.

HNTB Corporation

HydroQual Inc.

Insituform Technologies Inc.

Institute for Environmental Technology & Industry, Korea

Jacobson Helgoth Consultants Inc.

Jason Consultants Inc.

Jordan, Jones, & Goulding Inc.

KCI Technologies Inc.

Kelly & Weaver, P.C.

Kennedy/Jenks Consultants

Komline Sanderson Engineering Corporation

Lawler, Matusky & Skelly Engineers, LLP

Limno-Tech Inc.

Lombardo Associates Inc.

Malcolm Pirnie Inc.

McKim & Creed

MEC Analytical Systems Inc.

MWH

New England Organics

Odor & Corrosion Technology Consultants Inc. (OCTC)

ONDEO Degremont Inc.

Oswald Green, LLC

PA Government Services Inc.

Parametrix Inc.

Parsons

Post, Buckley, Schuh & Jernigan

R&D Engineering/Conestoga Rover & Associates

The RETEC Group

R.M. Towill Corporation

Ross & Associates Ltd.

Royce Technologies

SAIC Maritime Technical Group

Sear-Brown

Sierra Environmental Services

SYNAGRO

Tetra Tech Inc.

Trojan Technologies Inc.

URS Corporation

USFilter NATC

Vortechnics Inc.

Wade-Trim Inc.

Weston Solutions Inc.

Woodard & Curran

WRc/D&B, LLC

WWETCO, LLC

INDUSTRY

American Electric Power

ChevronTexaco Energy Research & Technology Company

The Coca-Cola Company

Dow Chemical Company

DuPont Company

Eastman Kodak Company

Merck & Company Inc.

ONDEO Services

Procter & Gamble Company

PSEG Services Corp.

Severn Trent Services Inc.

Shell Global Solutions

Thames Water Plc

United Water Services LLC

Note: List as of 5/12/03

WERF Product Order Form

As a benefit of joining the Water Environment Research Foundation, subscribers are entitled to receive one complimentary copy of all final reports and other products. Additional copies are available at cost (usually $10). To order your complimentary copy of a report, please write "free" in the unit price column. WERF keeps track of all orders. If the charge differs from what is shown here, we will call to confirm the total before processing.

Name ____________________ Title ____________________

Organization ____________________

Address ____________________

City ________ State ________ Zip Code ________ Country ________

Phone ________ Fax ________ Email ________

Stock #	Product	Quantity	Unit Price	Total
			Postage & Handling	
			VA Residents Add 4.5% Sales Tax	
			Canadian Residents Add 7% GST	
			TOTAL	

Method of Payment:

(All orders must be prepaid.)

❑ Check or Money Order Enclosed

❑ Visa ❑ Mastercard ❑ American Express

Account No. ____________________ Exp. Date ________

Signature ____________________

Shipping & Handling:

Amount of Order	United States	Canada & Mexico	All Others
Up to but not more than:	Add:	Add:	Add:
$20.00	$5.00*	$8.00	50% of amount
30.00	5.50	8.00	40% of amount
40.00	6.00	8.00	
50.00	6.50	14.00	
60.00	7.00	14.00	
80.00	8.00	14.00	
100.00	10.00	21.00	
150.00	12.50	28.00	
200.00	15.00	35.00	
More than $200.00	Add 20% of order	Add 20% of order	

*minimum amount for all orders

Note: Please make checks payable to the Water Environment Research Foundation.

To Order (Subscribers Only):

Log on to www.werf.org and click on the "Product Catalog."

Phone: (703) 684-2470
Fax: (703) 299-0742.

WERF
Attn: Subscriber Services
601 Wythe Street
Alexandria, VA 22314-1994

To Order (Non-Subscribers):

Non-subscribers may be able to order WERF publications either through WEF (www.wef.org) or IWAP (www.iwapublishing.com). Visit WERF's website at www.werf.org for details.

IWA Publishing's authorised EU representative for General Product Safety Regulations is Diane D'Arras, 15 rue Duret, 75116 Paris, France, e-mail: safety@iwap.co.uk.

Printed and bound by CPI Group (UK) Ltd, Croydon, CR0 4YY
12/05/2026
02108876-0013